Suggested Reading

Like many before and after me, my first substantive exposure to Gödel's incompleteness theorems came not by way of studying the famous 1931 paper itself but rather by reading, as an undergraduate, the celebrated *Gödel's Proof* by Ernest Nagel and James R. Newman (New York: New York University Press, 1968). This is a popular exposition that yet manages to go into some detail concerning the substance of the proof. My world was rocked. On rereading it after all these years, I was impressed all over again. It's a wonderful little book, in its own way a classic.

Jaakko Hintikka's very slim (70 pages) book, *On Gödel* (Belmont, CA: Wadsworth Thomson Learning, 2000), is also a clear and concise presentation of Gödel's proof for the non-expert. Like the more expansive *Gödel's Proof*, Hintikka's proof is self-contained, requiring no previous knowledge of logic. He also has a good sense of humor.

So far as the life of the logician is concerned, *Logical Dilemmas: The Life and Work of Kurt Gödel* (Wellesley, MA: A K Peters, 1997) by John Dawson is definitive. As not only a logician but also Gödel's archivist, whose wife learned to translate Gödel's shorthand, Dawson was in an unrivaled position for presenting the life of

Gödel. I was told by Institute mathematician Armand Borel that Gödel's literary remains, which had been donated to the Institute for Advanced Study by Gödel's widow, were in utter chaos, piled helter-skelter into decaying boxes; and then "a young man" (Dawson) had offered to put it all into order. "He did a good job, I'm told." Indeed he did.

John Dawson also has two papers on Gödel that are accessible and interesting: "Kurt Gödel in Sharper Focus" and "The Reception of Gödel's Incompleteness Theorems." Both are reprinted in *Gödel's Theorem in Focus,* edited by Stuart Shanker, as are other interesting essays, including Solomon Feferman's "Kurt Gödel: Conviction and Caution."

Hao Wang produced three rather eccentric but intriguing books out of the pickings of Gödel's mind: *From Mathematics to Philosophy* (New York: Humanities Press, 1974), *Reflections on Kurt Gödel* (Cambridge, MA: MIT Press, 1987), and *A Logical Journey* (Cambridge, MA: MIT Press, 1996). The books recount conversations Wang had with Gödel, interlaced with history of the logician's life and Wang's own views on the topics he and Gödel discussed. What they lack in structure they compensate for in content.

There are several memoirs of Gödel, written by those who had first known him in Vienna, and they are fascinating and in their own way touching. There is first of all Georg Kreisl's "Kurt Gödel: 1906-1978," *Biographical Memoirs of Fellows of the Royal Society*, Vol. 26 (1980), pp. 148–224. Kreisl, an eminent mathematical logician, is in a unique position, having known Wittgenstein quite well when Kreisl was a student, and then, later, having gotten to know Gödel in Princeton. Karl Menger had been invited, together with Gödel, to join the Vienna Circle as favored students of Hans Hahn and his invaluable first-hand reminiscences of Gödel are recounted in "Memories of Kurt Gödel," in *Reminiscences of the Vienna Circle and the Mathematical Colloquium*, ed. Louise Golland, Brian McGuinness, and Abe Sklar (Dordrecht: Kluwer, 1994). And then there is Olga

Taussky-Todd, herself a number-theorist, who also had first come to know Gödel in their student days. Her "Remembrances of Kurt Gödel" is in *Gödel Remembered* (Naples: Bibliopolis, 1987).

If the reader is interested in seeing how a contemporary polymath applies Gödel's theorems in his own creative scientific thinking, then he is advised to read Roger Penrose's *The Emperor's New Mind: Concerning Computers, Minds, and the Laws of Physics* (New York: Penguin, 1989) and his *Shadows of the Mind: A Search for the Missing Science of Consciousness* (Oxford: Oxford University Press, 1994). Like Gödel, Penrose is a confirmed mathematical Platonist; he interprets the incompleteness theorems exactly as Gödel did. There's lots of further fascinating mathematics that he discusses—including Turing's contributions to the work Gödel began, the Mandelbrot set, and Penrose's own work on the tiling of the plane—all argued, by him, as pointing in the direction of Platonism. Penrose's overall argument is that mathematical knowledge, the amazing fact that we *have* it, is evidence that the laws of physics are of a fundamentally different character than we have heretofore dreamt.

Douglas Hofstadter's Pulitzer-prize-winning *Gödel, Escher, Bach: The Eternal Golden Braid* (New York: Basic Books, 1974) is a spirited romp through self-referentiality. Hofstadter does a wonderful job of braiding together ideas in logic, art, and music, just as the title promises. When, upon being asked what I'd been working on these past few years, I'd say "Gödel," more often than not I got a blank stare in return. Then I'd mention the title of Hofstadter's bestseller, and the blank stare would give way to a smile and an "oh yes."

Raymond M. Smullyan is a mathematical logician who has written various exuberantly playful books that explore themes in logic, most especially self-referential paradoxes. His *Gödel's Incompleteness Theorems* (Oxford Logic Guides, no. 19, 1995) is written with his characteristic lucidity and verve, approaching the proofs from various angles and making use of the reformulations of Gödel's work that subsequent logicians, including Smullyan himself, have contributed. The book even comes with exercises that help the reader to assimilate the various subtleties of the proof, such as the strangeness of having arithmetical propositions that can talk to the condition of their own unprovability. No more logical background is required than a semester's course in symbolic logic or, failing that, a good book in formal logic, such as Smullyan's own *First-Order Logic* (Dover, 1992).

Finally, there is the writing of Gödel himself, his few published papers and his many unpublished works, in *Collected Works*, ed. Solomon Feferman et al. (Oxford: Oxford University Press, 1986–). There are five volumes to date.

Acknowledgments

If there is any aspect of being the perfect literary agent that Tina Bennett lacks, I have yet to discover it. The current project only served to reveal new aspects of Tina's ways of unstintingly supporting her writers.

I am extremely grateful to the following people who shared their recollections of Kurt Gödel with me: John Bahcall, Paul Benacerraf, Armand Borel, Thomas Nagel, Morton White. Each one was stintless with his time. Simon Kochen not only spoke long hours with me but also generously read over my completed manuscript, catching some technical errors, for which I am profoundly grateful, and answering further queries by e-mail. Berel Lang also read the manuscript and his comments, too, were insightful, substantive, and helpful.

I thank John Dawson, not only for the Herculean work he did as Gödel's archivist, which made the job of all scholars who follow possible, but also his prompt response to any question that arose.

As always, Sheldon Goldstein, physicist-philosopher, had insights that were invaluable. He, more than anyone, helped me to ease my way back into mathematical logic. There is not a man on Earth, I'd wager, quite equal to him for reminding one of the beauty and elegance of abstract thought. Steven Pinker generously read some early inchoate chapters, when I was feeling my way toward "popular technical writing," and his comments and encouragement were sustaining. And when I threw up my hands, Yael Goldstein calmly placed them back on the keyboard, offering the sort of sage advice and substantive criticism and guidance without which this book, quite literally, would not have been written.

Accordingly, I dedicate the book to her, with gratitude, love, and stupefied admiration.

Index

Praise for *Incompleteness: The Proof and Paradox of Kurt Gödel*

"Goldstein does a formidable job."

—Anthony Doerr, *Boston Globe*

"Beguilingly empathetic. . . . *Incompleteness* sticks in the mind like *Longitude*, another book about an enigmatic scientific genius who registers on the reader's mind as a mesmerizing void in human form. . . . Goldstein has a real knack for grounding even the loftiest theoretical disquisitions in reassuringly earthbound particulars. . . . To Goldstein's enduring credit . . . we come away from *Incompleteness* with a sense of Gödel both at his most brilliant and, later, at his most neurotic." —David Kipen, *San Francisco Chronicle*

"Magnificent. . . . Goldstein is an excellent choice for this installment of Norton's Great Discoveries series: Her philosophical background makes her a sure guide to the underlying ideas, and she brings a novelistic depth of character and atmosphere . . . to her sympathetic depiction of the logician's tortured psyche, as his relentless search for logical patterns . . . gradually darkened into paranoia." —*Publishers Weekly*

"Rebecca Goldstein has managed to get inside the head of Kurt Gödel and see what makes him tick. She presents

Gödel's mathematics as a consequence of his broader philosophical concerns. In her view Gödel is primarily a philosopher, but one who expressed his views through his mathematics. The strong connection between Gödel and Leibniz, which is little-known but was very important to Gödel, is particularly well documented. Highly recommended!"

—Gregory J. Chaitin,
author of *Meta Math!: The Quest for Omega*

"This is difficult material, at the borders of what we understand about human knowledge. The author has skillfully humanized it by showing us Gödel, Wittgenstein, and Einstein in their work, their friendships, and their disagreements. Perhaps only a novelist could have done this. Rebecca Goldstein has, in any case, done it superbly well."

—John Derbyshire, *New York Sun*

"[Goldstein] writes with a light touch that readers are sure to enjoy."
—Martin Davis, *Nature*

"*Incompleteness* is an artfully written and thoroughly engaging account of one of the greatest mathematical minds of the last century. By interweaving well-chosen episodes in Kurt Gödel's life with a detailed yet remarkably accessible account of his most stunning breakthrough—a proof that there are true but unprovable statements—Goldstein reveals both

Gödel's torment and his genius. By the book's end, we understand well why Einstein would look forward to 'the privilege of walking home with Gödel,' and we can't help but wish that we'd been able to join them." —Brian Greene, author of *The Elegant Universe* and *The Fabric of the Cosmos*

"In this penetrating, accessible, and beautifully written book, Rebecca Goldstein explores not only the work of one of the greatest mathematicians but also the relation of the human mind to the world around it."

—Alan Lightman, author of *Einstein's Dreams*

PUBLISHED TITLES IN THE GREAT DISCOVERIES SERIES

David Foster Wallace
Everything and More: A Compact History of ∞

Sherwin B. Nuland
The Doctors' Plague: Germs, Childbed Fever, and the Strange Story of Ignác Semmelweis

Michio Kaku
Einstein's Cosmos: How Albert Einstein's Vision Transformed Our Understanding of Space and Time

Barbara Goldsmith
Obsessive Genius: The Inner World of Marie Curie

Rebecca Goldstein
Incompleteness: The Proof and Paradox of Kurt Gödel

Madison Smartt Bell
Lavoisier in the Year One: The Birth of a New Science in an Age of Revolution

George Johnson
Miss Leavitt's Stars: The Untold Story of the Forgotten Woman Who Discovered How to Measure the Universe

David Leavitt
The Man Who Knew Too Much: Alan Turing and the Invention of the Computer

William T. Vollmann
Uncentering the Earth: Copernicus and The Revolutions of the Heavenly Spheres

FORTHCOMING TITLES

Richard Reeves on Rutherford and the Atom

Daniel Mendelsohn on Archimedes and the Science of the Ancient Greeks

David Quammen on Darwin and Evolution

General Editors: Edwin Barber and Jesse Cohen

BY REBECCA GOLDSTEIN

The Mind-Body Problem

The Late-Summer Passion of a Woman of Mind

The Dark Sister

Strange Attractors

Mazel

Properties of Light

Incompleteness

REBECCA GOLDSTEIN

Incompleteness

W. W. NORTON & COMPANY

NEW YORK • LONDON

Photograph Credits

p. 22 The American Institute of Physics/Emilio Segré Visual Archives
p. 253 Leonard McCombe/Timepix

First published as a Norton paperback 2006

Book design by Chris Welch
Production manager: Julia Druskin

Library of Congress Cataloging-in-Publication Data

Goldstein, Rebecca, date.
Incompleteness : the proof and paradox of Kurt Gödel / Rebecca Goldstein.— 1st ed.
p. cm. — (Great discoveries)
Summary: "An introduction to the life and thought of Kurt Gödel, who transformed our conception of math forever"—Provided by publisher.
Includes bibliographical references and index.
ISBN 0-393-05169-2
1. Gödel, Kurt. 2. Logicians—United States—Biography. 3. Logicians—Austria—Biography. 4. Proof theory. I. Title. II. Series.
QA29.G58G65 2005
510'.92—dc22

2004023052

ISBN 978-0-393-32760-1

W. W. Norton & Company, Inc., 500 Fifth Avenue, New York, N.Y. 10110
www.wwnorton.com

W. W. Norton & Company Ltd.
15 Carlisle Street, London W1D 3BS

For Yael
the child is mentor to the mother

Contents

But every error is due to extraneous factors (such as emotion and education); reason itself does not err.

—KURT GÖDEL

29 November 1972

Introduction

Exiles

It's late summer in suburban New Jersey. Down a secluded road two men are strolling, hands clasped behind their backs, quietly speaking. Above them a thick canopy of trees shelters them from the sky. Stately old homes stand far back from the road, while on the other side, just beyond the elms, the lush green carpet of a golf course rolls away, the muted voices of men at play coming as if from a great distance.

Yet, appearances to the contrary notwithstanding, this is not just one more suburban enclave strictly populated by the country club set, with men commuting daily into the city to support the affluence. No, this is Princeton, New Jersey, home of one of the great universities of the world, and so possessed of a far more eclectic population than a first glance suggests. At this moment that finds these two men strolling home on a quiet back road, Princeton's population has become even more cosmopolitan, with many of Europe's finest minds on the run from Hitler. As one American educator put it, "Hitler shakes the tree and I gather the apples."

Some of the choicest of apples have ended up rolling into this little corner of the world.

So it is not so surprising that the language in which the two strollers are conversing is German. One of the men, dapperly dressed in a white linen suit with a matching white fedora, is still in his thirties while the other, in baggy pants held up by old-world-style suspenders, is approaching seventy. Despite the difference in their ages, they seem to be talking to one another as peers, though occasionally the older man's face crinkles up into a well-worn matrix of amusement and he shakes his head as if the other has now said something *wirklich verrückt*, really cracked.

At one end of the leafy road, in the direction away from which the two are heading, the spanking new red-brick Georgian building of the Institute for Advanced Study is laid out on a great expanse of lawn. The Institute has been around now for over a decade, renting space in Princeton University's Gothic mathematics building. But the brainy influx from Europe has boosted the Institute's prestige, and now it has moved a few short miles from the university onto its own spacious campus, which includes a pond and a small forest, crisscrossed by paths, where fugitive ideas can be run to ground.

The Institute for Advanced Study is already, in the early 1940s, an American anomaly, peopled with a few select thinkers. Perhaps part of the explanation for the Institute's uniqueness lies in its having developed out of the visionary ideas of a single man. In 1930, educational reformer Abraham Flexner had persuaded two New Jersey department store heirs, Louis Bamberger and his sister Mrs. Felix Fuld, to charter a new type of academy, dedicated to the "usefulness of useless knowledge." The two retail magnates, motivated by their philanthropic intent, had sold their business to R. H. Macy and Co. just weeks before the stock market crash; with a fortune of $30 million, they had turned to Flexner to advise them on how to apply it to the betterment of mankind's mind.

Flexner, the son of Eastern European immigrants, had taken it upon himself some years before single-handedly to expose the shoddiness of American medical education. Around the turn of the century there was a surplus of medical schools, granting medical degrees that often indicated little more than that the recipient had paid the required tuition. The state of Missouri alone had 42 medical schools, the city of Chicago 14. Flexner's report, exposing the sham and published by the Carnegie Foundation for the Advancement of Teaching, had made a difference. Some of the worst of the institutions folded up their tents and snuck off into the night.

The Bamberger/Fulds were grateful to their former New Jersey patrons and wanted to give them something back. Their first thought was a medical school, and so they sent their representatives to speak with the man who knew so much about how medicine ought to be taught. (Flexner's brother was head of Rockefeller University's medical school, which served Flexner as a model.) But Flexner had been harboring even more utopian dreams than ensuring that American doctors know something about medicine. His thoughts on educational reform had taken a decided turn away from the applied and practical. His idea was to create a haven for the purest of thinkers, to realize the proverbial ivory tower in solid red brick: in short, to create what would come to be known as the Institute for Advanced Study.

Here the reverentially chosen faculty would be treated as

the princes of Reine Vernunft, of pure reason, that they were. They would be given generous remuneration (so that some dubbed the place "the Institute for Advanced Salaries"), as well as the priceless luxury of limitless time in which to think, unburdened of the need to prepare classroom lectures and correct student exam booklets—in fact unburdened of the presence of students altogether. Instead a constantly replenished stream of gifted younger scholars, eventually known as the "temporary members," would visit for one or two years, injecting the bracing tonic of their energy, youth, and enthusiasm into the ichor of genius. "It should be a free society of scholars," Flexner wrote. "Free, because mature persons, animated by intellectual purposes, must be left to pursue their own ends in their own way." It ought to provide simple, though spacious, surroundings "and above all tranquility—absence of distraction either by worldly concerns or by parental responsibility for an immature student body." The Bamberger/Fulds had originally wanted to locate their school in Newark, New Jersey, but Flexner persuaded them that Princeton, with its centuries-old traditions of scholarship and insulated layers of serenity, would be far more conducive to drawing forth the desired results from unfettered genius.

Flexner decided to establish his vision on the firm foundations of mathematics, "the severest of all disciplines," in his words. Mathematicians, in a certain sense, are the farthest removed of all academics from thoughts of "the real world"—a phrase which, in this context, means more than merely the practical world of current affairs. The phrase is meant to cover just about everything that physically exists, aside from ideas, concepts, theories: the world of the mind. Of course, the world of the mind can certainly be, and typically is, *about* the real world; however, not, typically, in mathematics. Mathematicians, in their extreme remoteness, may not enjoy (or suffer) much notice from the public at large; but, among those who live the life of the mind, they are regarded with a special sort of wonder for the rigor of their methods and the certainty of their conclusions, unique features that are connected with some of the very reasons that make them largely useless ("useless" in the sense that the knowledge of mathematics leads, in and of itself, to no practical consequences, no means of changing our material condition, for better or for worse).

The rigor and certainty of the mathematician is arrived at *a priori*, meaning that the mathematician neither resorts to any observations in arriving at his or her mathematical insights[1] nor do these mathematical insights, in and of themselves, entail observations, so that nothing we experience can undermine the grounds we have for knowing them. No experience would count as grounds for revising, for example, that 5 + 7 = 12. Were we to add up 5 things and 7 things, and get 13 things, we would recount. Should we still, after repeated recounting, get 13 things we would assume that one of the 12 things had split or that we were seeing double or dreaming or

1 This is not, however, to imply that these beliefs are innate, i.e., that we are born having them. Obviously, we must first acquire the concepts, and the language for expressing them, before we can come to believe that 5 + 7 = 12. Innateness is a psychological notion, whereas aprioricity is an epistemological notion, having to do with the way in which the belief is justified, what counts as evidence both for and against it. [Note: Two types of notes will be employed in this book: footnotes to continue a thought on the page, endnotes to give citations.]

even going mad. The truth that 5 + 7 = 12 is used to evaluate counting experiences, not the other way round.

The a priori nature of mathematics is a complicated, confusing sort of a thing. It's what makes mathematics so conclusive, so incorrigible: Once proved, a theorem is immune from empirical revision. There is, in general, a sort of invulnerability that's conferred on mathematics, precisely because it's a priori. In the vaulting tower of Reine Vernunft the mathematicians stand supreme on the topmost turret, their methods consisting of thinking, and thinking alone; this is partly what Flexner meant by calling their discipline the most severe.

Despite their intellectual stature, mathematicians are relatively cost-effective to maintain, requiring, again in Flexner's words, only "a few men, a few students, a few rooms, books, blackboards, chalk, paper, and pencils." No expensive laboratories, observatories, or heavy equipment is required. Mathematicians carry all their gear in their craniums, which is another way of saying that mathematics is a priori. Calculated also in Flexner's practical reasoning was the fact that mathematics is one of the few disciplines in which there is almost total unanimity on the identity of the best. Just as mathematics, alone among the disciplines, is able to establish its conclusions with the unassailable finality of a priori reason, so, too, the ranking of its practitioners follows with almost mathematical certitude. Flexner, functioning not only as the Institute's designer but as its first director, would know exactly after whom to go.

He soon loosened the requirements sufficiently to allow for the most theoretical of physicists and mathematical of economists, and late in 1932 was able to make the triumphant announcement that the first two employees he had hired were Princeton's own Oswald Veblen, a mathematician of the highest rank, and none other than Germany's Albert Einstein, the scientist whose near-cult status had made him a prominent target for the Nazis. Einstein's revolutionary theories of special and general relativity had been attacked by German scientists as representative of pathologically "Jewish physics," corrupted by the Jewish infatuation with abstract mathematics. Even before the genocidal plans kicked into full operation, the physicist had been placed on the Third Reich's special hit list.

As would be expected, a host of universities were more than willing to open their doors to so prestigious a refugee; in particular, Pasadena's California Institute of Technology was vigorously trying to recruit him. But Einstein favored Princeton, some say because it was the first American university to show interest in his work. His friends, casting their cosmopolitan eyes at the New Jersey seat of learning, convinced of its essential provinciality, asked him "Do you want to commit suicide?" But with a homeland suddenly turned maniacally hostile, perhaps Princeton's early and lasting friendliness proved irresistible. Einstein asked Flexner for a salary of $3000 and Flexner countered with $16,000. Soon the famous head with the ion-charged hair was strolling the suburban sidewalks, so that at least on one occasion a car hit a tree "after its driver suddenly recognized the face of the beautiful old man walking along the street."

Other luminaries from Europe followed Einstein to New Jersey, including the dazzling Hungarian polymath, John von Neumann, who would begin construction of the world's first computer while at the Institute, scandalizing those members who shared Flexner's commitment to keeping the Institute free of any "useful" work.[2] But it is Albert Einstein who has been immortalized, even while still very much alive,[3] as the apotheosis of the man of genius, so that townspeople have, almost since the day of his arrival, taken to calling Flexner's establishment "the Einstein Institute."

Sure enough the older of the two strollers glimpsed on the leafy road leading from the Institute is none other than Princeton's most famous denizen, his face once again registering wry amusement at something his walking partner has just propounded in all apparent seriousness. The younger man, a mathematical logician, acknowledges Einstein's reaction by producing a faint, crooked smile of his own, but continues to deduce the implications of his idea with unflappable precision.

The topics of their daily conversations range over physics and mathematics, philosophy and politics, and in all of these areas the logician is more than likely to say something to startle Einstein in its originality or profundity, naïveté or downright outlandishness. All of his thinking is governed by an "interesting axiom," as Ernst Gabor Straus, Einstein's assistant from 1944 to

2 As the first venture of the Institute outside the realm of purely theoretical work, it was criticized as "out of place" even by faculty members who had a high regard for the endeavor itself, according to the official account of the Institute's School of Mathematics. After von Neumann's death, the computer was quietly transferred to Princeton University.

3 Many contemporaries report the "awed hush" (in the words of Helen Dukas, ibid.) that would fall over a lecture or seminar room when he entered. Princeton philosopher Paul Benacerraf, who had been a graduate student at Princeton in Einstein's day, told me that Einstein sometimes used to attend the weekly Friday philosophy seminar, seldom speaking but still making his presence felt simply because it was his presence.

1947, once characterized it. For every fact, there exists an explanation as to why that fact *is* a fact; why it *has* to be a fact. This conviction amounts to the assertion that there is no brute contingency in the world, no givens that need *not* have been given. In other words, the world will never, not even once, speak to us in the way that an exasperated parent will speak to her fractious adolescent: "*Why?* I'll tell you *why*. Because I said so!" The world always has an explanation for itself, or as Einstein's walking partner puts it, *Die Welt ist vernünftig*, the world is intelligible. The conclusions that emanate from the rigorously consistent application of this "interesting axiom" to every subject that crosses the logician's mind—from the relationship between the body and the soul to global politics to the very local politics of the Institute for Advanced Study itself—often and radically diverge from the opinions of common sense. Such divergence, however, counts as nothing for him. It is as if one of the unwritten laws of his thought processes is: If reasoning and common sense should diverge, then . . . so much the worse for common sense! What, in the long run, *is* common sense, other than common?

This younger man is known to far fewer people, in his own day as well as in ours. Yet his work was, in its own way, as revolutionary as Einstein's, to be grouped among the small set of the last century's most radical and rigorous discoveries, all with consequences seeming to spill far beyond their respective fields, percolating down into our most basic preconceptions. At least within the mathematical sciences, the first third of the twentieth century made almost a habit of producing conceptual revolutions. This man's theorem is the third leg, together with Heisenberg's uncertainty principle and Einstein's relativity, of that tripod of theoretical cataclysms that have been felt to force disturbances deep down in the foundations of the "exact sciences." The three discoveries appear to deliver us into an unfamiliar world, one so at odds with our previous assumptions and intuitions that, nearly a century on, we are still struggling to make out where, exactly, we have landed.

It is much in the remote nature both of the man and of his work that he will never approach the celebrity status of his Princeton walking partner nor of the author of the uncertainty principle, who is almost certainly engaged at this same moment in history in the effort to produce the atom bomb for Nazi Germany. Einstein's walking partner is a revolutionary with a hidden face. He is the most famous mathematician

The logician and the physicist on one of their daily walks to and from the Institute for Advanced Study, Princeton.

that you have most likely never heard of. Or if you have heard of him, then there is a good chance that, through no fault of your own, you associate him with the sorts of ideas—subversively hostile to the enterprises of rationality, objectivity, truth—that he not only vehemently rejected but thought he had conclusively, *mathematically*, discredited.

He is Kurt Gödel, and in 1930, when he was 23, he had produced an extraordinary proof in mathematical logic for something called the incompleteness theorem—actually two logically related incompleteness theorems.

Unlike most mathematical results, Gödel's incompleteness theorems are expressed using no numbers or other symbolic formalisms. Though the nitty-gritty details of the proof are formidably technical, the proof's overall strategy, delightfully, is not. The two conclusions that emerge at the end of all the formal pyrotechnics are rendered in more or less plain English. The *Encyclopedia of Philosophy*'s article "Gödel's Theorem" opens with a crisp statement of the two theorems:

> By Gödel's theorem the following statement is generally meant:
>
> In any formal system adequate for number theory there exists an undecidable formula—that is, a formula that is not provable and whose negation is not provable. (This statement is occasionally referred to as Gödel's first theorem.)
>
> A corollary to the theorem is that the consistency of a formal system adequate for number theory cannot be proved within the system. (Sometimes it is this corollary that is referred to as Gödel's theorem; it is also referred to as Gödel's second theorem.)

> These statements are somewhat vaguely formulated generalizations of results published in 1931 by Kurt Gödel then in Vienna. ("Über formal unentscheidbare Sätze der Principia Mathematica und verwandter Systeme I," received for publication on November 17, 1930.)

Though one might not guess it from this terse statement of them, the incompleteness theorems are extraordinary for (among other reasons) how *much* they have to say. They belong to the branch of mathematics known as formal logic or mathematical logic, a field which was viewed, prior to Gödel's achievement, as mathematically suspect;[4] yet they range far beyond their narrow formal domain, addressing such vast and messy issues as the nature of truth and knowledge and certainty. Because our human nature is intimately involved in the discussion of these issues—after all, in speaking of knowledge we are implicitly speaking of knowers—Gödel's theorems have also seemed to have important things to say about what our minds could—and could not—be.

Some thinkers have seen in Gödel's theorems high-grade grist for the postmodern mill, pulverizing the old absolutist ways of thinking about truth and certainty, objectivity and rationality. One writer expressed the postmodern sentiment in lively eschatological terms: "He [Gödel] is the devil, for math. After Gödel, the idea that mathematics was not just a

4 Before Gödel came onto the scene, logicians were more likely to be members of a philosophy department. Simon Kochen, a logician in the mathematics department at Princeton University, remarked to me that "Gödel put logic on the mathematical map. Every mathematical department of note now has logic represented on its staff. It may only be one or two logicians, but there will, at least, be someone" (May 2002).

language of God but language we could decode to understand the universe and understand everything—that just doesn't work any more. It's part of the great postmodern uncertainty that we live in." The necessary incompleteness of even our formal systems of thought demonstrates that there is no non-shifting foundation on which any system rests. All truths—even those that had seemed so certain as to be immune to the very possibility of revision—are essentially manufactured. Indeed the very notion of the objectively true is a socially constructed myth. Our knowing minds are not embedded in truth. Rather the entire notion of truth is embedded in our minds, which are themselves the unwitting lackeys of organizational forms of influence. Epistemology is nothing more than the sociology of power. So goes, more or less, the postmodern version of Gödel.

Other thinkers have argued that, in regard to the nature of the human mind, the implications of Gödel's theorems point in an entirely different direction. For example, Roger Penrose in his two bestselling books, *The Emperor's New Mind* and *Shadows of the Mind*, has made the incompleteness theorems central to his argument that our minds, whatever they are, cannot be digital computers. What Gödel's theorems prove, he argues, is that even in our most technical, rule-bound thinking—that is, mathematics—we are engaging in truth-discovering processes that can't be reduced to the mechanical procedures programmed into computers. Notice that Penrose's argument, in direct opposition to the postmodern interpretation of the previous paragraph, understands Gödel's results to have left our mathematical knowledge largely intact. Gödel's theorems don't demonstrate the limits of the human mind, but rather the limits of *computational* models of the human mind (basically, models that reduce all thinking to rule-following). They don't leave us stranded in postmodern uncertainty but rather negate a particular reductive theory of the mind.

Gödel's theorems, then, appear to be that rarest of rare creatures: mathematical truths that also address themselves—however ambiguously and controversially—to the central question of the humanities: what is involved in our being human? They are the most prolix theorems in the history of mathematics. Though there is disagreement about precisely how much, and precisely what, they say, there is no doubt that they say an awful lot and that what they say extends beyond mathematics, certainly into metamathematics and perhaps even beyond. In fact, the metamathematical nature of the theorems is intimately linked with the fact that the *Encyclopedia of Philosophy* stated them in (more or less) plain English. The concepts of "formal system," "undecidable," and "consistency" might be semi-technical and require explication (which is why the reader should not worry if the succinct statement of the theorems yielded little understanding); but they are metamathematical concepts whose explication (which will eventually come) is *not* rendered in the language of mathematics. Gödel's conclusions are mathematical theorems that manage to escape mere mathematics. They speak from both inside *and* outside mathematics. This is yet another facet of their distinct fascination, the facet seized upon in yet another popular book, Douglas Hofstadter's Pulitzer-prize-winning *Gödel, Escher, Bach: An Eternal Golden Braid.*

The prefix *meta* comes from the Greek, and it means "after," "beyond," suggesting the view from outside, as it were. The metaview of a cognitive area poses such questions as: how is it possible for this area of knowledge to be doing what it is doing? Mathematics, just because it is sui generis—the severest of disciplines—using a priori methods to establish its often astounding, though incorrigible, results, has always forcefully presented theorists of knowledge (known as "epistemologists") with metaquestions, most specifically, the question of how it is possible for it to be doing what it is doing. The certainty of mathematics, the godlike infallibility it seems to bestow on its knowers, has been seen as presenting both a paradigm to be emulated—if we can do it there, let's do it everywhere[5]—and also a riddle to be pondered: how *can* we do it, there or anywhere? How can the likes of us, thrown up out of the blindfolded thrashings of evolution, attain any sort of infallibility? To grasp this riddle it might be helpful to recall a famous remark of Groucho Marx's, to the effect that he would not belong to any country club that would accept the likes of him. Similarly, some have fretted that if mathematics is really so certain then how can it be known by the likes of us? How can we have gained entry into so restricted a cognitive club?

Metaquestions about a field, say about science or mathematics or the law, are not normally questions that are contained in the field itself; they are not, respectively, scientific or

5 This utopian epistemology is characteristic of the seventeenth-century rationalists—René Descartes (1596–1650), Benedictus Spinoza (1632–1677), and Gottfried Wilhelm Leibniz (1646–1716). Spinoza and Leibniz, in particular, believed it was possible to appropriate the standards and methods of the mathematicians and generalize them so that they could answer all our posed questions: scientific, ethical, even theological. Then, when theological differences of the sort that cause long and bloody wars arose, men of reason could respond: "Come, let us *a priori* deduce."

mathematical or legal. Rather they are categorized as philosophical questions, residing, respectively, in the philosophy of science, of mathematics, of law. Gödel's theorems are spectacular exceptions to this general rule. They are at once mathematical and metamathematical. They have all the rigor of something that is a priori proved, and yet they establish a metaconclusion. It is as if someone has painted a picture that manages to answer the basic questions of aesthetics; a landscape or portrait that represents the general nature of beauty and perhaps even explains why it moves us the way it does. It is extraordinary that a mathematical result should have anything at all to say about the nature of mathematical truth in general.

Gödel's two theorems address themselves to the very issue that has always singled out mathematics: the certainty, the incorrigibility, the aprioricity. Do the theorems cast us out of the most exclusive cognitive club in epistemology, undermining our claim of being able to attain, in the area of mathematics at least, perfect certitude? Or do the theorems leave us members in good standing? Gödel himself, as we shall see, held strong convictions on this metaquestion, sharply at odds with interpretations that are commonly linked with his work.

For both Gödel and Einstein, metaquestions of how, respectively, physics and mathematics are to be interpreted—what it is that these powerful forms of knowledge actually do and how they do it—are central to their technical work. Einstein, too, had extremely strong metaconvictions regarding physics. More specifically, Einstein's and Gödel's metaconvictions were addressed to the question of whether their respective fields are descriptions of an objective reality—existing independent of our thinking of it—or, rather, are subjective human projections, socially shared intellectual constructs.

The emphasis that each placed on these metaquestions was, in itself, enough to separate them from most practitioners in their respective fields. Not only were both men centrally interested in the metalevel, but, even more unusually, they also wanted their technical work to shed metalight. Gödel, in fact, had acquired the ambition, while still an undergraduate at the University of Vienna, of devoting himself *only* to the sort of mathematics that would have broader philosophical implications. This is a truly daunting goal, in some sense historically ambitious, and one of the most astounding aspects of his story is that he managed to achieve it. This daunting ambition, which he preserved throughout his life, may have limited how much he did, but it also determined that whatever he did was profound. Einstein, though not quite so strict with himself as Gödel, still shared the conviction that truly good science always keeps the larger philosophical questions in view: "Science without epistemology is—insofar as it is thinkable at all—primitive and muddled."

The friendship between Einstein and Gödel is still the stuff of both legend and speculation. Every day the two men made the trek back and forth from the Institute, and others watched them with curiosity and wondered that they had so much to say to one another. For example, Ernst Gabor Straus wrote:

> No story of Einstein in Princeton would be complete without mentioning his really warm and very close friendship with Kurt Gödel. They were very, very dissimilar people, but for some reason they understood each other well and appreciated each other enormously. Einstein often mentioned that he felt that he should not become a mathematician because the wealth of interesting and attractive

problems was so great that you could get lost in it without ever coming up with anything of genuine importance. In physics, he could see what the important problems were and could, by strength of character and stubbornness, pursue them. But he told me once, "Now that I've met Gödel, I know that the same thing does exist in mathematics." Of course, Gödel had an interesting axiom by which he looked at the world; namely, that nothing that happens in it is due to accident or stupidity. If you really take that axiom seriously all the strange theories that Gödel believed in become absolutely necessary. I tried several times to challenge him, but there was no out. I mean, from Gödel's axioms they all followed. Einstein did not really mind it, in fact thought it quite amusing. Except the last time we saw him in 1953, he said, "You know, Gödel has really gone completely crazy." And so I said, "Well, what worse could he have done?" And Einstein said, "He voted for Eisenhower."

Straus's language indicates a certain puzzlement as to what the two men saw in one another; in particular, what the sagacious physicist could have seen in the neurotic logician. Einstein, wrote Straus, was "gregarious, happy, full of laughter and common sense." Gödel, on the other hand was "extremely solemn, very serious, quite solitary and distrustful of common sense as a means of arriving at the truth."

The Einstein of legend—with his wild hair and absentmindedness, his quixotic embrace of one-world politics and other lost causes—is not usually portrayed as a savvy, worldly sort; but, compared to Gödel, he was. Most in Princeton, even his mathematical colleagues, found Gödel, with his "interesting axiom" exponentially complicating every discussion and practical decision, all but impossible to speak with. As the mathematician Armand Borel wrote in his history of the Institute's School of Mathematics, he and the others sometimes "found the logic of Aristotle's successor. . . quite baffling." Eventually, the mathematicians solved their Gödel problem by banishing him from their meetings, making him a department of one: the sole decision-maker on anything having strictly to do with logic.

Though Princeton's population is well accustomed to eccentricity, trained not to look askance at rumpled specimens staring vacantly (or seemingly vacantly) off into space-time, Kurt Gödel struck almost everyone as seriously strange, presenting a formidable challenge to conversational exchange. A reticent person, Gödel, when he *did* speak, was more than likely to say something to which no possible response seemed forthcoming:

John Bahcall was a promising young astrophysicist when he was introduced to Gödel at a small Institute dinner. He identified himself as a physicist, to which Gödel's curt response was "I don't believe in natural science."

The philosopher Thomas Nagel recalled also being seated next to Gödel at a small gathering for dinner at the Institute and discussing the mind-body problem with him, a philosophical chestnut that both men had tried to crack. Nagel pointed out to Gödel that Gödel's extreme dualist view (according to which souls and bodies have quite separate existences, linking up with one another at birth to conjoin in a sort of partnership that is severed upon death) seems hard to reconcile with the theory of evolution. Gödel professed himself a nonbeliever in evolution and topped this off by pointing out, as if this were additional corroboration for his own rejection of Darwinism: "You know Stalin didn't believe in evolution either, and he was a very intelligent man."

"After that," Nagel told me with a small laugh, "I just gave up."[6]

The linguist Noam Chomsky, too, reported being stopped dead in his linguistic tracks by the logician. Chomsky asked him what he was currently working on, and received an answer that probably nobody since the seventeenth-century's Leibniz had given: "I am trying to prove that the laws of nature are a priori."

Three magnificent minds, as at home in the world of pure ideas as anyone on this planet, yet they (and there are more) reported hitting an insurmountable impasse in discussing ideas with Gödel.

Einstein, too, was presented time and again, on their daily walks to and from the Institute, with examples of Gödel's strange intuitions, his profound "anti-empiricism." Nevertheless Einstein consistently sought out the logician's company. In fact,

6 Gödel's hostility to the theory of evolution becomes quite understandable the more one understands his mind. A rationalist like Gödel wishes to excise chance and randomness, whereas natural selection invokes randomness and contingency as fundamental explanatory factors. At the level of microevolution (generation-to-generation changes), the theory gives a central role to random mutation and recombination. At the level of macroevolution (patterns in the history of life), it gives a central role to historical contingency, such as the vagaries of geology and climate, or such chance events as a meteorite's crashing to Earth, blackening out the Sun, wiping out the dinosaurs, thus allowing mouselike mammals to inhabit the vacated ecological niches. (I am indebted to Steven Pinker for this insight.)

economist Oskar Morgenstern,[7] who had known Gödel back in Vienna, confided in a letter: "Einstein had often told me that in the late years of his life he has continually sought Gödel's company, in order to have discussions with him. Once he said to me that his own work no longer meant much, that he came to the Institute merely *um das Privileg zu haben, mit Gödel zu Fuss nach Hause gehen zu dürfen*," that is, in order to have the privilege of walking home with Gödel. Even given their shared interest in the metalevel of their respective fields, Einstein's avowal of devotion strikes one as extravagant.

For his part, Gödel's letters to his mother, Marianne, who remained behind in Europe (a correspondence that gives us some knowledge of his life until her death in 1966), are filled with references to Einstein. If Einstein, in his last years, went to the Institute merely for the privilege of walking home with Gödel, for Gödel there was simply nobody else in all the world with whom to talk, at least not in the way in which he could talk to Einstein (an exclusivity made all the more poignant when one considers that Gödel had a wife). So that, for example, on 4 July 1947, he wrote to his mother that Einstein had been ordered by his doctor to take a rest cure. "So I am now quite lonesome and speak scarcely with anybody in private."

It was, and remains, a minor mystery to those who observed their powerful friendship. "I used to see them walking across the path from Fuld Hall to Olden Farm every day," the Swiss-born

7 Morgenstern had also fled Nazi-occupied Austria for the Institute. Even though he was an economist, his work was sufficiently mathematical—he is one of the founders, with von Neumann, of game theory—to gain him entrance into Flexner's Institute.

Armand Borel, who came to the Institute a little after Gödel, told me as I sat in his office at the Institute. "I do not know what it was they spoke about. It was most probably physics, because Gödel, too, was interested in physics, you know.[8] They didn't want to speak to anybody else. They only wanted to speak to each other," he concluded with a shrug.

It is important in understanding the relationship between Einstein and Gödel, in trying to peer behind Straus's bemused "somehow they understood each other very well," not simply to stop short at the easy explanation that these two were uniquely each other's intellectual peer, that they constituted, in the logician Hao Wang's words, a "two-membered 'natural kind' consisting of the leading 'natural philosophers' of the century."[9] There is much more, even beyond membership in so exclusive a set, to be said for what bound the two together.

There are the surface similarities, of course. There is the fact, for example, that they had both done their most important work in Central Europe, in German-speaking lands, from which they had been forced to flee. But in this respect at least, Einstein and Gödel were hardly unique in the Princeton of their day. Scholar after scholar had had to flee Vienna and Göttingen and Budapest for places like Pasadena and Princeton. The fact that they were political exiles, who spoke the same native tongue and found themselves strolling the

8 Gödel produced a very original solution to the field equations of Einstein's general relativity, and surprised Einstein with them for his seventieth birthday. In Gödel's solution, time is cyclical. See Chapter 4.

9 Hao Wang (1921–1995), a logician at Rockefeller University, devoted himself to understandiing Gödel's views on everything from the nature of mathematical intuition to the transmigration of souls, and produced three books out of the material.

improbable landscape of suburban New Jersey, certainly does not begin to explain the special bond between them, which mystified even their fellow refugees.

There are other striking similarities between the two. There is, for example, the fact that both of them had done their most important work when quite young men. Einstein had been 26 in 1905, his *annus mirabilis*, when, as an obscure patent clerk in Bern, Switzerland, he had published his articles on (special) relativity, the light quantum, and Brownian motion, as well as completed his Ph.D. dissertation. Gödel's results (which also were three in number, though it is the first incompleteness theorem that far outshines all else[10]) had been accomplished three years before reaching the comparable age.

More important than this shared autobiographical detail is the fact that each man had toyed, at an even earlier age, with the idea of entering the field that the other had chosen. Gödel had entered the University of Vienna intending to study physics. Einstein had first thought of becoming a mathematician. There is a sense in which each saw in the other a realization of what he might have become had he opted otherwise, and there was undoubtedly a certain fascination in this.

Still there is far more that bound the two of them together. I would like to propose that the reason for the profound understanding and appreciation that held between these two "very, very dissimilar people" lay on the deepest level of their revolutionary ideas. They were comrades in the most profound sense in which thinkers can be comrades. Both men were committed to an understanding of reality, and of their own work in rela-

10 His other two achievements dating 1929–30 were the second incompleteness theorem and the proof of the completeness of the predicate calculus.

tion to that reality, that placed them painfully at odds with the international community of thinkers.

One might have thought, with each having presented results so uniquely transformative that their respective fields had been forced to *remake* themselves to contain these results at their center, that the last thing that Einstein and Gödel would have felt is marginalized. Feelings of alienation, disaffection, dismissal, isolation are for the noninfluential and the failed. But disaffected and even dismissed they felt, and, moreover, disaffected and dismissed in profoundly similar ways, at the metalevel of their fields, the level at which you interpret *what it all means.*

There is a sense, then, at least as I have tried to penetrate to the core of a friendship that mystified onlookers, in which Einstein and Gödel were fellow exiles within a larger exile, and it is a sense that goes far beyond the geopolitical conditions that caused them to seek safety in Princeton, New Jersey. I believe that they were fellow exiles in the deepest sense in which it is possible for a thinker to be an exile. Strange as it might seem for men so celebrated for their contributions, they were intellectual exiles.

To fully understand their sense of shared isolation, which provided the cohesive force to their famous friendship, it will be necessary to consider the metaconvictions that alienated them from their peers. How ought we to interpret, in terms of the larger philosophical questions, Einstein's relativity theory and Gödel's incompleteness theorems? How did the authors of these masterpieces of human thought interpret them and how did others?

Gödel's incompleteness theorems. Einstein's relativity theories. Heisenberg's uncertainty principle. The very names are tantalizingly suggestive, seeming to inject the softer human element into the hard sciences, seeming, even, to suggest that the human element *prevails* over those severely precise systems, mathematics and theoretical physics, smudging them over with our very own vagueness and subjectivity. The embrace of subjectivity over objectivity—of the "nothing-is-but-thinking-makes-it-so" or "man-is-the-measure-of-all-things" modes of reasoning—is a decided, even dominant, strain of thought in the twentieth-century's intellectual and cultural life. The work of Gödel and Einstein—acknowledged by all as revolutionary and dubbed with those suggestive names—is commonly grouped, together with Heisenberg's uncertainty principle, as among the most compelling reasons modern thought has given us to reject the "myth of objectivity." This interpretation of the triadic grouping is itself part of the modern—or, more accurately, postmodern—mythology.

So, for example, in the 1998 acclaimed play *Copenhagen*, the playwright Michael Frayn not only correctly presents the physicists Niels Bohr and Werner Heisenberg as rejecting the idea that physics is descriptive of an objective physical reality, but he also inaccurately identifies Einstein's relativity theory as the first of modern physics' moves in the direction of that ultimate rejection:

> Bohr: It [quantum mechanics] works, yes. But it's more important than that. Because you see what we did in those three years, Heisenberg? Not to exaggerate but we turned the world inside out. Yes, listen, now it comes, now it comes. . . . We put man back at the center of the universe. Throughout history we keep finding ourselves displaced. We keep exiling ourselves to the periphery of things. First

we turn ourselves into a mere adjunct of God's unknowable purposes. Tiny figures kneeling in the great cathedral of creation. And no sooner have we recovered ourselves in the Renaissance, no sooner has man become, as Protagoras proclaimed him, the measure of all things, than we're pushed aside again by the products of our reasoning! We're dwarfed again as physicists build the great new cathedrals for us to wonder at—the laws of classical mechanics that predate us from the beginning of eternity, that will survive us to eternity's end, that exist whether we exist or not. Until we come to the beginning of the twentieth century, and we're suddenly forced to rise from our knees again.

Heisenberg: It starts with Einstein.

Bohr: It starts with Einstein. He shows that measurement—measurement, on which the whole possibility of science depends—measurement is not an impersonal event that occurs with impartial universality. It's a human act, carried out from a specific point of view in time and space, from the one particular viewpoint of a possible observer. Then here in Copenhagen in those three years in the mid-twenties we discover that there is no precisely determinable objective universe. That the universe exists only as a series of approximations. Only within the limits determined by our relationship with it. Only through the understanding lodged inside the human head.

Like Einstein's relativity theory, Gödel's incompleteness theorems have been seen as holding a prominent place in the twentieth-century's intellectual revolt against objectivity and rationality. For example, in a popular work of philosophy written by William Barrett, *Irrational Man: A Study in Existentialist Philosophy,* published in 1962 while Gödel was still alive (and which I was required to read the summer before entering college), Gödel is placed alongside such thinkers as Martin Heidegger (1889–1976) and Friedrich Nietzsche (1844–1900), destroyers of our illusions of rationality and objectivity:

> Gödel's findings seem to have even more far-reaching consequences [than Heisenberg's Uncertainty and Bohr's Complementarity], when one considers that in the Western tradition, from the Pythagoreans and Plato onward, mathematics as the very model of intelligibility has been the central citadel of rationalism: Now it turns out that even in his most precise science—in the province where his reason had seemed omnipotent—man cannot escape his essential finitude; every system of mathematics that he constructs is doomed to incompleteness. Gödel has shown that mathematics has insoluble problems, and hence can never be formalized in any complete system. . . . Mathematicians now know they can never reach rock bottom; in fact, there is no rock bottom, since mathematics has no self-subsistent reality independent of the human activity that mathematicians carry on.

Barrett correctly states the (first) incompleteness theorem, that mathematics can never be formalized in any complete system. And the philosophical conclusion he draws from it is very much in sync with the most fashionable intellectual trends of the twentieth century. So it might surprise the reader to learn that Gödel himself drew no such conclusion. In fact, if we replace the "no" before "self-subsistent reality" with an "a," we will arrive at an accurate statement of Gödel's own metamathematical view, the view that inspired all of his mathematical work, including his famous incompleteness theorems.

Though intellectual gurus may have interpreted Gödel as falling into step with the great revolt against objectivity and rationality that characterizes much of twentieth-century thinking, this was not the interpretation that Gödel himself held of his revolutionary results. Precisely the same thing can be truthfully said of Einstein. Both men, in fact, were staunch believers in objectivity and interpreted their own most famous work as support positive of this increasingly unpopular position. While so many of their intellectual peers might have made the subjectivist turn—citing the great achievements of relativity theory and the incompleteness theorems as signposts pointing them in that direction—Einstein and Gödel did not.

Both Einstein and Gödel are as far from seconding the ancient Sophist's "man is the measure of all things" as it is perhaps possible to be. For both of these men the methodology of their respective fields—the complex mixtures of reasoning, including both intuition and deduction (and, in the case of physics, which is not a priori, observation as well)—does not consist of arbitrary sets of rules that govern an elaborate made-up mind-game or language-game, which could just as well have been played by some other sets of rules entirely, leading to an altogether different construction of reality. No, for both thinkers these are the rules that lead our minds out beyond the circumscriptions of personal experience to gain access to aspects of reality that it is impossible otherwise to know.

Einstein's profound isolation from his scientific peers is as well known (if as little understood) as most other aspects of his celebrated life. It is often explained as stemming from his curmudgeonly refusal to accept the revolutionary advance of quantum mechanics, in particular its fundamentally stochastic nature, from which the element of pure chance cannot be excised. The familiar story told about him is that, having made his own conceptual revolution as a young man with his relativity theories, both special and general, he settled, as is the wont of older men, into a conservative mind-set unable to wrap itself around the revolutions of the next generation, even if those later revolutions were the logical extensions of his own. This telling of Einstein's story is also part of the intellectual mythology of the twentieth century.

Yet it is not accurate. The heart of Einstein's scientific alienation is his *rejection* of the subjectivist turn that the playwright has his characters declare "all began with Einstein." Einstein had not understood his relativity theory as pointing toward the subjectivist interpretation of physics but, rather, precisely in the opposite direction. "Relativity," as it occurs in Einstein's theory, means something far more technical and restricted than that measurement (and so everything) is relative to human points of view.[11] For Einstein, in fact, to have

11 Measurements of properties like length are, according to special relativity, relative to a particular coordinate system or reference frame. But to reduce these technical terms—coordinate system, reference frame—to the idea of human points of view, is, well, nonsense. We have the choice of various coordinate systems to describe the motion of something, and, according to the theory of relativity, all coordinate systems are equal; none is privileged. In one coordinate system an "observer" (who does not even have to be a *conscious* entity, and hence does not have to be literally observing or even *capable* of observing anything) will be at rest; in another he or she or it will be moving. It's often natural, though not determined, to choose a coordinate system with respect to which a particular observer is at rest. Thus it is often natural (though not determined) to choose the coordinate system in which the earth, for example, is at rest. The motion of all of us terrestrials, with our myriad subjective points of view would then be described relative to *one* coordinate system, in which the earth is at rest.

followed men like Werner Heisenberg and Niels Bohr in the direction of subjectivity would have been to deny what he took to be the most fundamental meta-implications of relativity theory. Einstein interpreted his theory as representing the *objective* nature of space-time, so very *different* from our human, subjective point of view of space and time.[12] Far from restoring us to the center of the universe, describing everything as relative to our experiential point of view, Einstein's theory, expressed in terms of beautiful mathematics, offers us a glimpse of an utterly surprising physical reality, surprising precisely *because* it is nothing like what we are presented with in our experiential apprehension of it.

Einstein sometimes speaks of objective reality as the "out yonder," and in the "Autobiographical Notes" that he supplied with his typical self-mocking good humor for the *Festschrift* that P. A. Schilpp edited in honor of the physicist's seventieth birthday,[13] he explicitly identifies his belief in this reality as the spiritual center of his life as a scientist:

> It is quite clear to me that the religious paradise of youth, which was thus lost, was a first attempt to free myself from the chains of the "merely personal," from an existence which is dominated by wishes, hopes, and primitive feel-

12 In relativity theory, for example, time doesn't flow, but rather is, as the fourth dimension, as static as space. In vivid contrast, the most dramatic (and poignant) aspect of our *experience* of time is its ceaseless, unidirectional motion, carrying us away from the past and toward the future.

13 "Here I sit in order to write, at the age of 67, something like my own obituary. I am doing this not merely because Dr. Schilpp has persuaded me to do it; but because I, in fact, believe that it is a good thing to show those who are striving alongside of us how one's striving and searching appears to one in retrospect" (Schilpp, p. 3).

> ings. Out yonder there was this huge world, which exists independently of us human beings and which stands before us like a great, eternal riddle, at least partially accessible to our inspection and thinking. The contemplation of this world beckoned like a liberation. . . . The mental grasp of this extra-personal world within the frame of the given possibilities swam as highest aim half consciously and half unconsciously before my mind's eye. . . . The road to this paradise was not as comfortable and alluring as the road to the religious paradise; but it has proved itself as trustworthy, and I have never regretted having chosen it.

This is an eloquent statement of Einstein's credo as a scientist, and it really could not be more at odds with the sentiments of almost all the other prominent physicists of his circle.[14] Einstein understood the business of physics to be to discover theories that offer a glimpse of the objective nature that stands "out yonder" behind our experiences. Werner Heisenberg, together with such men as the Danish Niels Bohr and the German Max Born (who are together the leading advocates of the Copenhagen interpretation of quantum mechanics) reject this view in the name of an intellectual movement known as "positivism," according to which any attempt to reach out beyond our experience results in arrant nonsense.

We will have occasion to look more closely at positivism in the next chapter, when we move from Princeton, New Jersey, to Vienna, Austria, and examine the circumstances that brought

14 Contrast it, for example, with this statement of Werner Heisenberg's: "The idea of an objective real world whose smallest parts exist objectively in the same sense as stones or trees exist, independently of whether or not we observe them . . . is impossible."

forth, almost as an act of supreme intellectual rebellion *against* the positivists, Gödel's two incompleteness theorems.

Positivism, especially as it came to be espoused by the group of scientists, mathematicians, and philosophers of the famed Vienna Circle, under the strong influence of the charismatic Viennese-born philosopher Ludwig Wittgenstein, is a severe theory of meaning that makes liberal use of the word *meaningless*. In particular, it brands as meaningless any descriptive proposition[15] that cannot in principle be verified through the contents of our experience. The meaning of a proposition is given by the means of empirically verifying it (the verificationist criterion of meaning).

Gödel, like Einstein, is committed to the possibility of reaching out, *pace* the positivists, beyond our experiences to describe the world "out yonder." Only since Gödel's field is mathematics, the "out yonder" in which he is interested is the domain of abstract reality. His commitment to the objective existence of mathematical reality is the view known as conceptual, or mathematical, realism. It is also known as mathematical Platonism, in honor of the ancient Greek philosopher whose own metaphysics was a vehement rejection of the Sophist Protagoras' "man is the measure of all things."

Platonism is the view that the truths of mathematics are independent of any human activities, such as the construction

15 By *descriptive proposition* one means a proposition that is not just true (or false) by virtue of its meaning alone. Propositions whose truth-value (truth or falsity) is a function of their meaning alone are called "analytic" or, sometimes, "trivial." So, for example, "all bilingual people speak at least two languages" is analytic. A proposition that is, on the other hand, descriptive is not true or false simply by virtue of its meaning but also by virtue of the facts of the matter. So the proposition "I am bilingual" is false by virtue of both its meaning and the facts of the matter.

of formal systems—with their axioms, definitions, rules of inference, and proofs. The truths of mathematics are determined, according to Platonism, by the reality of mathematics, by the nature of the real, though abstract, entities (numbers, sets, etc.) that make up that reality. The structure of, say, the natural numbers (which are the regular old counting numbers: 1, 2, 3, etc.) exists independent of us, according to the mathematical realist, just as does the structure of space-time, according to the physical realist; and the properties of the numbers 4 and 25—that, for example, one is even, the other is odd and both are perfect squares—are as objective as are, according to the physical realist, the physical properties of light and gravity.

For Gödel mathematics is a means of unveiling the features of objective mathematical reality, just as for Einstein physics is a means of unveiling aspects of objective physical reality. Gödel's understanding of what we are doing when we are doing mathematics could be rendered in words echoing Einstein's credo: "Out yonder there is this huge world, which exists independently of us human beings and which stands before us like a great, eternal riddle, at least partially accessible to our inspection and thinking." Only here the "out yonder" is to be understood as at an even further remove from the subject of experience, with his distinctly human point of view. The "out yonder" is out beyond physical space-time; it is a reality of pure abstraction, of universal and necessary truths, and our faculty of a priori reason provides us—mysteriously—with the means of accessing this ultimate "out yonder," of gaining at least partial glimpses of what might be called (in the current fashion in naming television shows: "Extreme Survival," "Extreme Makeover," "The Most Extreme") "*extreme* reality."

Gödel's mathematical Platonism was not in itself unusual. Many mathematicians have been mathematical realists; and even those who do not describe themselves as such, when they are cornered and asked pointblank about their metamathematical position, will slip unself-consciously into realism when they speak of their work as their "discoveries."[16] G. H. Hardy (1877–1947), an English mathematician of great distinction, expressed his own Platonist convictions in his classic *A Mathematician's Apology*, with no apologies at all:

> I believe that mathematical reality lies outside of us, that our function is to discover or *observe* it, and that the theorems which we prove, and which we describe grandiloquently as our "creations," are simply our notes of our observations. This view has been held, in one form or another, by many philosophers of high reputation from Plato onwards, and I shall use the language which is natural to a man who holds it. . . .
>
> [T]his realistic view is much more plausible of mathematical than of physical reality, because mathematical objects are so much more what they seem. A chair or a star is not in the least like what it seems to be; the more we think of it, the fuzzier its outlines become in the haze of sensation which surrounds it; but "2" or "317" has nothing to do with sensation, and its properties stand out the more clearly the more closely we scrutinize it. It may be that modern physics fits best into some framework of idealistic philosophy—I do not believe it, but there are eminent physicists who say so. Pure mathematics, on the other hand, seems to me a rock on which all idealism founders: 317 is a prime, not because we think so, or because our

16 Interestingly, this is true even of David Hilbert, whose formalism was sharply opposed to Platonism (see chapter 2).

minds are shaped in one way or another, but *because it is so*, because mathematical reality is built that way.[17]

The almost three millennia since Plato have given us plenty of new and amazing mathematics, but not much more reason to believe in Platonism than the ancient Greek philosopher himself had. Mathematician after mathematician has testified, like Hardy, to their Platonist conviction that they are discovering, rather than creating, mathematical truths. However, testifying is just about all we ever got . . . until Gödel. Gödel's audacious ambition to arrive at a mathematical conclusion that would simultaneously be a metamathematical result supporting mathematical realism was precisely what yielded his incompleteness theorems.

Gödel's metamathematical view, his affirmation of the objective, independent existence of mathematical reality, constituted perhaps the essence of his life, which is to say what is undoubtedly true: that he was a strange man indeed. His philosophical outlook was not an expression of his mathematics; his mathematics were an expression of his philosophical outlook, his Platonism, which was the deepest expression,

17 The circumstances of the writing of Hardy's classic are as moving as they are unusual. Hardy had lost his mathematical creativity, which tends to happen to mathematicians relatively young. (A mathematician of 40 has probably already seen his best years, which is why the most prestigious award for mathematicians [there is no Nobel Prize for mathematics], the Fields Medal, is awarded to someone 40 or younger.) Hardy attempted suicide, survived the attempt, and was persuaded by C. P. Snow to write a book explaining the life of a mathematician. The result, *A Mathematician's Apology*, is incomparable. Soon after completing it, Hardy again attempted suicide, and succeeded.

therefore, of the man himself. That his work, like Einstein's, has been interpreted as not only consistent with the revolt against objectivity but also as among its most compelling driving forces is then more than a little ironic.

Einstein was fortunate in his last years to have a kindred philosophical spirit, even if one as unstable and fey as Gödel, to soften the sense of exile. The words that Morgenstern quoted from Einstein, that in his last years he went to his Institute office only in order to have the privilege of walking home with Gödel, become, in the subtleties of the metalight, less surprising.

After Einstein's death in 1955, Gödel's sense of intellectual exile deepened; his most profound sense of identification was with the über-rationalist Leibniz, who had been dead for almost 300 years. The explanations the logician arrived at through the rigorous application of his "interesting axiom" took on ever darker tones. The young man in the dapper white suit shriveled into an emaciated man, entombed in a heavy overcoat and scarf even in New Jersey's hot humid summers, seeing plots everywhere. He came to believe that there was a vast conspiracy, apparently in place for centuries, to suppress the truth "and make men stupid." Those who had discovered the full power of a priori reason, men such as the seventeenth-century's Leibniz and the twentieth-century's Gödel, were, he believed, marked men. His profound isolation, even alienation, from his peers provided fertile soil for that rationality run amuck which is paranoia.

That the greatest logician since Aristotle should have followed reason so unwaveringly to such illogical conclusions has struck many people as paradoxical. But, as I hope will become ever clearer in the chapters to come, the internal paradoxes in Gödel's personality were at least partially provoked by the world's paradoxical responses to his famous work. His incompleteness theorems were simultaneously celebrated and ignored. Their technical content transformed the fields of logic and mathematics; the method of proof he used, the concepts he defined in the course of the proof, led to entirely new areas of research, such as recursion theory and model theory. Other central areas of research were abandoned, particularly those sanctioned by the greatest mathematician of the generation just preceding Gödel's, David Hilbert (1862–1943), having been shown to be futile by reason of Gödel's theorems.

Yet the metamathematical import of the theorems, which to Gödel was their most important aspect, was disregarded. Even more paradoxically, the racier currents in the culture, hawking postmodern uncertainty and the false mythology of all absolutes, scooped his theorems up, together with Einstein's relativity, reinterpreting them so that they precisely negated the convictions that Gödel and his fellow exile had so passionately wanted to demonstrate.

Paradoxes, in the technical sense, are those catastrophes of reason whereby the mind is compelled by logic itself to draw contradictory conclusions. Many are of the self-referential variety; troubles arise because some linguistic item—a description, a sentence—potentially refers to itself. The most ancient of these paradoxes is known as the "liar's paradox," its lineage going back to the ancient Greeks.[18] It is centered on

18 Here is the textual reference the paradox is derived from: "One of themselves, even a prophet of their own, said, 'The Cretans are always liars.' . . . This witness is true" (Titus 1:12–13).

the self-referential sentence: "This very sentence is false." This sentence must be, like all sentences, either true or false. But if it is true, then it is false, since that is what it says; and if it is false, well then, it is true, since, again, that is what it says. It must, then, be both true *and* false, and *that* is a severe problem. The mind crashes.

Paradoxes like the liar's play a technical role in the proof that Gödel devised for his extraordinary first incompleteness theorem. Gödel was able to take the structure of self-referential paradoxicality—the sort of structure that causes our minds to crash when considering "This very sentence is false"—and turn it into an extraordinary proof for one of the most surprising results in the history of mathematics.[19] This itself seems almost paradoxical. Paradoxes have always seemed specifically designed to convince us that we are simply not smart enough to take up whatever topic brought us to them. Gödel was able to twist the intelligence-mortifying material of paradox into a proof that leads us to deep insights into the nature of truth, and knowledge, and certainty. According to Gödel's own Platonist understanding of his

19 That mathematical conclusions have the ability to surprise us might itself seem paradoxical. The world might very well, and often does, confound our expectations, our experiential contact with it bringing us to rude awakenings. But how can conclusions that are arrived at through purely a priori reason do so? If a priori truths are—by definition—immune from empirical revision, then it's not some unexpected experience of the world that delivers the punch. We ourselves must deduce the confoundment, and this seems prima facie odd. This metamathematical issue, too, is addressed by Gödel's prolix first theorem. For Gödel, the independent reality of mathematics, of which our axioms are only incomplete descriptions, takes the surprise out of the surprisingness of mathematics.

proof, it shows us that our minds, in knowing mathematics, are escaping the limitations of man-made systems, grasping the independent truths of abstract reality.

The structure of Gödel's proof, the use it makes of ancient paradox, speaks at some level, if only metaphorically, to the paradoxes in the tale that the twentieth century told itself about some of its greatest intellectual achievements—including, of course, Gödel's incompleteness theorems. Perhaps someday a historian of ideas will explain the subjectivist turn taken by so many of the last century's most influential thinkers, including not only philosophers but hard-core scientists, such as Heisenberg and Bohr. Such an explanation lies well beyond the scope of this book. But what I can do is to describe the effects that the revolt against objectivity had on one of the twentieth century's greatest thinkers: how it provoked him into his proof of the incompleteness theorems and how it then reinterpreted those theorems as confirmation of itself.

To understand the full richness—and paradox—of Gödel, his world and his work, it will be necessary to take two steps backward from the glimpse of him walking home with Einstein on a shady road in Princeton. We'll step back first to the 1920s Vienna of his youth, the scene of so many of the young century's intellectual and cultural assaults on tradition; and then take another retreat back to the turn of the century, when a conception of mathematics gave birth to a program for completing mathematics that would fall victim to the work of the reticent young logician with the outsized metamathematical ambitions.

I
A Platonist among the Positivists

First Love

Kurt Gödel was 18 when he arrived in Vienna to begin his studies at the university. Though he had been born in Moravia, in what is now the Czech Republic but was then part of the Hapsburg Empire, his arrival in Vienna must have felt like something of a homecoming. He had considered himself an exile even in the land of his birth.

He was born on 28 April 1906 in Brno, or what the Germans and Austrians still call "Brünn." His parents, Rudolf and Marianne, were of German rather than Czech origin, and associated exclusively with the other Sudeten Germans who dominated in Brno. The city was the center of the Hapsburg Empire's textile industry,[1] so when Rudolf proved to be no scholar at grammar school, he was enrolled, at the age of 12, at a weaver's

1 It was also, interestingly, the location of the Augustinian monastery where Gregor Mendel (1822–1884) performed his highly tedious and important experiments with pea plants, resulting in his discovery of the laws of dominance and recessiveness in heredity.

school, where he found his calling. He completed his studies there with distinction and was given a job in the textile factory of Friedrich Redlich, where he worked until his death. He rose swiftly through the ranks, eventually becoming a director and joint partner. Consequently, the family lived comfortably, eventually acquiring a villa in a fashionable neighborhood.

Gödel's mother, Marianne, was far more educated and cultured than his father, which was not unusual among the bourgeoisie of the Empire. It was also common for marriage choices to be forged out of practical concerns rather than romantic inclinations, and this, too, seemed to be the case in the Gödels' marriage. As so often happens in such cases, the mother's strongest emotional ties were supplied by her children, in her case Rudolf, born a year after her marriage, and then four years later Kurt, who was baptized Kurt Friedrich, the middle name honoring his father's employer, who served as godfather. For some reason, the logician dropped his middle name when he became a U.S. citizen in 1948.

Almost all of our knowledge of Kurt Gödel's earliest years, as sparse as it is, comes by way of his older brother Rudolf, who wrote a brief "History of the Gödel Family," as well as from Rudolf's responses to queries from the logicians Hao Wang and John Dawson on the subject of his younger brother's childhood. (Rudolf was a physician who never married and remained in Austria. He died in 1992, at the age of 90.)

We learn from Rudolf that Kurt asked so many questions that his nickname was *der Herr Warum*, or Mr. Why. Little children, as anyone who has spent any significant amount of time with them knows, tend to push the "why" questions pretty hard. We are born into a sort of ontological wonder (*thaulamazein*) that passes into oblivion as we get used to the

The Gödel family, ca. 1910: Marianne, Kurt, father Rudolf, brother Rudolf.

lay of the land. Gödel's intense childhood *thaulamazein* persisted throughout his adult life, so that the child who was called *Herr Warum* grew into the man who began the 14 principles of his private credo with *Die Welt is vernünftig*: the world is rational. Like many gifted mathematicians, Gödel reached a certain level of precocious maturity while still a young child; then, having arrived at this level, he remained there. The picture *en famille* we have of the future "successor to Aristotle"[2] at age four shows a cherubic little man, seriously

2 Aristotle is commonly acknowledged as the father of logic. His work in logic is laid out in the *Prior Analytics,* which is part of the posthumous consortium known as the *Organon.* The philosopher had the seminal

staring straight into the camera, his hand precisely poised before him, the little forward hunch giving the suggestion of solemn contemplation.

We also learn from Rudolf, in a letter to the logician Hao Wang, that at about the age of five, the younger brother suffered a mild anxiety neurosis ("*leichte Angst Neurose*"), and at the age of eight he suffered a severe bout of "joint rheumatism, with high fever." The patient did research on his illness and, learning that the illness could cause possible permanent heart damage, he inferred that precisely this outcome had occurred in his particular case. Gödel held to the conviction of an injured heart throughout his life, despite the absence of any evidence. The conclusion he reached as an eight-year-old child, entirely on his own, was to contribute to his lifelong hypochondria.

When the random permutations of genetic blending produce an offspring whose intelligence far outstrips that of his parents that child faces a special sort of predicament: he both recognizes his utter dependence, being after all only a child; and he also clearly perceives the severe limits of his own par-

---

insight that in a deductive logical argument, some words are logically relevant while others are not. The irrelevant words can be dispensed with by making them variables. So, for example, in the stock syllogism: *if all men are mortal, and Socrates is a man, then Socrates is mortal*, the words "men" and "mortal," and "Socrates" are disposable. This particular syllogism is just an instantiation of the more general syllogism-scheme: *if all X's are Y, and i is an X, then i is a Y.* The move toward denoting logically irrelevant words with variables was a move toward generality and thus toward the science of logic. Aristotle, however, generalized too much, asserting that all deductive reasoning is syllogistic. The modern developments of the nineteenth century, most especially those of the German Gottlob Frege (1848–1925), revealed the greater variety of deductive arguments.

ents' understanding. Most people come to the latter recognition only during adolescence, when the normal reaction is an explosive mixture of hubris, contempt, and outrage (how can they be so dumb?). But the reaction of a young child is more likely to be blind terror (how can they be trusted to take care of me?). The *leichte Angst Neurose* is some indication that the precocious Gödel grasped the limits of parental omniscience at about the age of five. It would be comforting, in the presence of such a shattering conclusion, especially when it's reinforced by a serious illness a few years later, to derive the following additional conclusion: There are always logical explanations and I am exactly the sort of person who can discover such explanations. The grownups around me may be a sorry lot, but luckily I don't need to depend on them. I can figure out everything for myself. The world is thoroughly logical and so is my mind—a perfect fit.

Quite possibly the young Gödel had some such thoughts to quell the terror of discovering at too young an age that he was far more intelligent than his parents. It would explain much about the man he would become. The child is father to the man—even more so, perhaps, in the case of mathematical geniuses.

At school, the K.-K. *Staatsrealgymnasium mit deutscher Unterrichtssprache*[3] (obviously, a German-language school), Gödel excelled in all his studies and began to show the

3 "K.-K." stands for *Kaiserliche-Königliche* (imperial-royal) and referred to all pertaining to the Austrian crownlands. *Kaiserliche und königliche* was applied to that which was jointly administered by Austria and Hungary, and *königliche* alone referred to that pertaining to Hungary. This system of imperial abbreviation seems ready-made for satirizing, and so it was, magisterially, by Robert Musil in his novel, *Man Without Qualities.*

marked aloofness and solemnity that would characterize him throughout his life. A fellow schoolmate, Harry Klepetař, wrote to John Dawson that "from the beginning . . . Gödel kept more or less to himself and devoted most of his time to his studies." He also reported that Gödel's interests were "manifold," and that "his interest in mathematics and physics [had already] manifested itself . . . at the age of 10."

Gödel, however, declined to take any courses in the language of the republic in which he was living (the native language of most of the students was German). Klepetař recalled to Dawson that Gödel was the only one of his fellow students he never heard speak a word of Czech and that, especially after October 1918 when the Czechoslovak Republic declared its independence, "Gödel considered himself always Austrian and an exile in Czechoslovakia." So the sense of exile began early in his life and, in its various senses, some more pernicious than others, it is doubtful that it ever left him.

Gödel entered the University of Vienna in 1924, intending to study physics, but, as he later told Hao Wang, "his interest in precision led him from physics to mathematics and to mathematical logic." Gödel's interest in physics had begun at the age of 15, when he read Goethe's theory of colors, which was embedded in a general attack on Newtonian physics. His transition to mathematics was also encouraged, he told Hao Wang, by the excellent teaching of some of his professors at the university. The course on number theory, given by Professor Phillip Furtwängler, attracted such huge numbers of students (up to 400) that it was necessary to issue alternate-day seating passes. Gödel was one of these rapt students, and he later said that they were the most wonderful lectures he had ever heard.

Intending to concentrate on number theory, he switched his major to mathematics in 1926, but in 1928 he began to work in mathematical logic. He was already a committed Platonist in 1926 when he turned from physics to mathematics. His metaphysical commitment had been forged the year before, when he took a course in the history of philosophy with Professor Heinrich Gomperz, whose father, Theodore, was a distinguished professor of ancient philosophy.

It is no easy task to penetrate the inner life of Kurt Gödel. One knows enough to recognize that it is markedly different from those of others, so reasoning by analogy will get us only so far. Then, too, he was the most reticent of men, assertively nondemonstrative in all things other than mathematics—where "demonstration" means, of course, something quite unique. He was a man of deep passions, as his life will bear out; but these passions were kept scrupulously hidden and they were rigorously intellectual.

I think it is fair to say, however, that like so many of us Gödel fell in love while an undergraduate. He underwent love's ecstatic transfiguration, its radical reordering of priorities, giving life a new focus and meaning. One is not quite the same person as before.

Kurt Gödel fell in love with Platonism, and he was not quite the same person as he was before.

What is the evidence for so transformative a passion roiling within the opaquely self-contained logician? That is, what is the evidence, in addition to the incompleteness theorems themselves?

Some of the evidence lies in the *Nachlass*, Gödel's literary remains, which are housed in Princeton's Firestone Library. The *Nachlass* had been left to molder in the Institute's basement, until John Dawson undertook the formidable task of becoming Gödel's archivist. (Gödel used a sort of shorthand script, Gabelsberger, that he had learned in high school, so the job involved, on top of everything else, translation.) Gödel had apparently kept almost every scrap of paper that had ever intercepted his life. There are journal articles, clothing bills, manuscripts, family pictures, student exercises, library slips for books he had borrowed in Vienna and Princeton. I found (in the Institute's collection of Gödeliana) those little Bible studies published by the Jehovah's Witnesses, the kind that their itinerants will urge on you if you happen to be home in the middle of the day and answer the door. These contained careful underlinings and marginalia in the logician's hand.

More telling are the many drafts of letters that were never posted, the manuscripts for articles that he had promised to deliver, that he had labored over with revision upon painstaking revision, and then never released for publication. One receives the impression of a cautiousness of hysterical proportions. Not a cautiousness of cerebration, for Gödel's intellectual ambitions were audacious, his intuitions fierce, his willingness to carry them to their logical conclusion undeterable. Rather there was a hysteria of discretion in presenting his thoughts to the external world.

Among the documents over which Gödel labored, and then never delivered, are the responses he made to a questionnaire that had been prepared for him by a sociologist. Burke D. Grandjean had made repeated attempts to interview Gödel and finally devised a questionnaire for him in 1974. (This is two years before the logician's death.) There are two slightly different versions of the completed questionnaire in the

Nachlass, as well as a typed, unsigned, and unsent letter addressed to Mr. Grandjean, dated 19 August 1975, and which begins in a rather bristling manner: "Dear Mr. Grandjean: Replying to your inquiries I would like to say first that I *don't* consider my work 'a facet of the intellectual atmosphere of the early 20th century' but rather the opposite." One can imagine the sociologist's reverential letter that had prompted this testy retort. In the context of so reticent a life, this letter, together with the two sets of replies to the questionnaire, is revealing. Gödel's bristling tone in his unsent response to Grandjean strengthens the sense that his life, especially after the death of Einstein, was characterized by a profound sense of intellectual isolation—a sense of isolation deepened by misinterpretations of his famous result.

Grandjean had listed various thinkers and asked Gödel to indicate which ones had influenced him, and Gödel made clear how far from the mark Grandjean's assumptions were. There seems to be a lifetime of exasperation behind the responses. Leibniz is not even listed.

To Grandjean's question: "Are there any influences to which you attribute special significance in the development of your philosophy?" Gödel's entire answer consisted of: "Heinrich Gomp. [erz] Prof[essor] of Ph[ilosophy] of Vienna." A strange answer, but, then again, not. It was in Professor Gomperz's class that Gödel's transfigurative intellectual love had been engendered. Though Gödel may have sat rapt in Professor Furtwängler's class on number theory, it was in Professor Gomperz's Introduction to the History of Philosophy that the true rapture transpired.

Plato has always had a strong appeal to the mathematically inclined. Plato himself was mathematically inclined. Written over the entrance to the Academy, the Athenian school of higher education he founded (essentially, the first European university), were the words: "Let no one enter herein who has not first studied geometry."

The ancient Greek philosopher's disdain for the Sophists, particularly for such men as Protagoras, gave the negative connotation to the word for these itinerant teachers. (The root of the word *Sophist* is the ancient Greek word for knowledge. *Philosophy*, literally "love of knowledge," shares the root.) Protagoras had meant his assertion that "man is the measure of all things" to apply most directly to the moral sphere; he had been arguing for what we now call "moral relativism," the claim that there is no objective difference between right and wrong, only different opinions, relativized either to individuals or to conglomerates of individuals who roughly share the same values (i.e., to societies). "True," when attached to moral opinions is an abbreviation for "true for *x*," where *x* is an individual or a society of ethically like-minded individuals.

Plato took on the relativists. It was his lifelong occupation. He not only argued for the objectivity of moral truth but he also founded his claim of objective truth—in the moral as well as other spheres—on his assertion of the objectivity of an abstract reality, graspable not through the senses but through reason.

The one area in which Platonism has proved most stubbornly resilient is mathematics, or rather metamathematics. A mathematician's sense that he is discovering objective truths, rather than simply constructing systems, is a commitment to Platonism. The conviction that such things as numbers and sets serve as models for our systems, which systems are true only insofar as they describe the nature of such things as numbers and sets, is likewise a commitment to Platonism.

First exposure to Plato can be an extremely heady experience for those with a passion for abstraction. (I remember my own.) It can amount to a sort of ecstasy. Plato himself argued that the beauty of the abstract realm, which immeasurably exceeds that of any single particular, can and ought to kindle a passion far larger than any prompted by individual beautiful persons (fickle, imperfect creatures who cannot even be counted on to love us in return and whose beauty not only cannot compete with the transcendent sort but also is subject to the corrosive actions of time). To have used the expression "fell in love" in relation to Gödel's undergraduate experience is to echo Plato himself, who used the most erotically charged language to describe the mind's approach toward and possession of the beauties of abstract objectivity.

> "Here is the life, Socrates, my friend," said the Manitean visitor, "that a human being should live—studying the beautiful itself. Should you ever see it, it will not seem to you to be on the level of gold, clothing, and beautiful boys and youths, who so astound you now when you look at them that you and many others are eager to gaze upon your darlings, and be together with them all the time. You would cease eating and drinking, if that were possible, and instead just look at them and be with them. What do we think it would be like," she said, "if someone should happen to see the beautiful itself, pure, clear, unmixed, and not contaminated with human flesh and color and a lot of other mortal silliness, but rather if he were able to look upon the divine, uniform, beautiful itself. Do you think," she continued, "it would be a worthless life for a human being to look at that, to study it in the required way, and be together with it?"

A "symposium" actually meant a drinking party in Plato's Athens, and in the dialogue to which Plato wryly gave that name he urges us to leave off lesser intoxications, including those associated with the sensual love of beautiful young things, and to become drunk on the beauty of truth—the sort of necessary and immutable truth acquired through pure reason, for which mathematics serves as the model. An aspect of the Platonic vision is a rejection of the easy bifurcation between passion, on the one side, and reason, on the other. Plato is urging us toward impassioned reason, the higher intoxication. Of course, susceptibility to the higher intoxication is predicated on the ability to grasp the intellectual love object, the beauties of pure abstraction, "to look at that, to study it in the required way, and be together with it." The young Kurt Gödel was singularly susceptible.

Gödel's reaction to the headiness of Plato's rapturous vision of truth was, it seems, the resolve to devote himself only to mathematics of (to recall Einstein's phrase) "genuine importance." It would have to be mathematics that had metasignificance, that was philosophically porous so that the objective source of all abstract truth could be seen to shine.

At first, Gödel had been drawn toward number theory because he believed that it would provide the strongest evidence for, and the clearest application of, conceptual realism. It was in 1928, when he was 22, that his mathematical interests began to shift toward mathematical logic. The fact that he had been attracted to number theory precisely because of his Platonist commitment, as he told Hao Wang, and that he then veered toward mathematical logic is tantalizing. Hao Wang did not ask the follow-up question. Exactly when did Gödel glimpse that logic might yield the metamathematical conclusions that he was seeking?

It is tempting to speculate about this, and informed speculation is the most that we have. We have no good idea of the path that led him to his theorems, by way of an ingenious form of argument the likes of which had never before been seen.

In contrast, we know a great deal about the preoccupations that had led Einstein to his special theory of relativity. It is all part of the public record of the scientist who performed the role of the professional genius in the collective imagination of the world. We know how, beginning at the age of 16, he used to perform *Gedanken*-experiments, thought-experiments, imagining himself hitching a ride on a light beam, or running along beside it, trying to deduce how the laws of physics would look from the point of view of an observer moving at the speed of light.

But Gödel's genius was never put on public display the way Einstein's was. The sources of his inspiration, the play of mind, revealing how ancient paradox could be transformed into a proof for conclusions shot through with meta-overtones, are unknown. He must somehow have glimpsed the metamathematical potential of logic, even when logic was, as it was then, far less mathematically respectable than his own work would render it. We do not know exactly when he proved his first incompleteness result. Not even his dissertation advisor (he had by then advanced to doing graduate work) knew what he was up to. But we do know that by 7 October 1930 he had the proof for the first incompleteness theorem.

The logician Jaakko Hintikka wrote:

> It is a measure of Gödel's status that the most important moment of his career is the most important moment in the history of twentieth-century logic, maybe in the history of logic in general. This *Sternstunde* was October 7, 1930. The setting was a conference on the foundations of mathematics in Königsberg on October 5–7, 1930.

What happened at Königsberg on 7 October 1930 was that Kurt Gödel, a relatively unknown graduate student attending a conference on metamathematics dominated by the leaders in the field, dropped a parsimonious few words indicating that he had a proof for the incompleteness of arithmetic. He was basically ignored by everyone present, with the exception of one mathematician, who happened to be there to represent a metamathematical position deeply at odds with Gödel's Platonism but who was astute enough to draw for himself the implications of Gödel's wildly muted "announcement."

Yet there is something wrong with what Hintikka says. The most important moment in Gödel's career did not come in the public revelation of the first incompleteness theorem. That moment just seems like the most important because that is when Gödel gave some slight public indication of what he had been up to. The most important moments of his career were, in fact, those about which we know nothing: the moments of intuitions or thought-experiments or God-knows-what that brought him to the proof itself.

His Platonist conviction must have convinced him, sans proof, that mathematical reality must exceed all formal attempts to contain it; but how did he lay hands on the strategy by which to prove incompleteness? How did it occur to him, in particular, to transform the structural features of self-referential paradoxes into a proof? How did the inspired idea of what we now call "Gödel numbering" come to him, the technique by means of which statements of mathematics would acquire double-entendres, making metamathematical statements as well? The overall strategy of the proof is astoundingly simple, the details that had to be worked out are astoundingly complicated, and both astounding features make us wish we knew more about how he came up with it all. But all we have is the result: the proof that forever changed our understanding of mathematics, and, in doing so, perhaps helped to change our understanding of ourselves.

So it is not Königsberg that is the scene of the real drama but rather Vienna—the Vienna of the late twenties and early thirties, a city utterly unique in its cultural and intellectual aspects. No thinker reflects in an utter vacuum—not even the purest of pure mathematicians up there on the topmost turret of Reine Vernunft. Not even a thinker so unwaveringly loyal to the integrity of his own intuitions as Kurt Gödel is utterly indifferent, if even in the spirit of opposition, to the prevailing opinions of his day, to the sorts of questions floating like spores in the intellectual atmosphere.

The city of Vienna in that period between the two world wars, the strange intensity of the thinking and the creating that were pursued there, plays its role in the story of Gödel's theorems. Vienna was then a city with a hugely disproportionate number of seminal thinkers and artists—scientists, musicians, poets, visual artists, philosophers, architects—who collectively seemed drawn into one sustained and intense conversation, pursued across every discipline and art form. Gödel, as reticent as he was, also ended up participating in this conversation.

Here, in this highly dramatic city, in which even intellectual life achieved a theatricality, even Gödel, the last person in the world to seek outward drama, attained a certain degree of it.

Out from the Muddle of the Old: A City in Search of New Foundations

If Princeton is a high-energy intellectual vortex disguising itself as a pleasantly bland spot on the suburban New Jersey landscape, the Vienna of the 1920s, when Kurt Gödel arrived there as a student, was "the research laboratory for world destruction" in the famous words of one of its contemporary chroniclers, the journalist and satirist Karl Kraus. The novelist Herman Kesten saw it as a city of "brilliant creation in a nonetheless decaying culture."

The teeming intellectual life of the city was carried out in broad sight, not only in university lecture halls and professors' offices but also in the numerous cafés that seem still to display the essence of Viennese life. Much had changed in the city, and the country as a whole, after the First World War and the collapse of the Hapsburg Empire in 1916. But Vienna remained, in its feel, a small large city—the undisputed cultural center of its country. The sense that it was almost an entirely self-enclosed entity within the greater country, sharing little in the way of outlook with the rest of the population, has perhaps a counterpart in contemporary New York City's relationship with the United States, though the discontinuity between city and general culture seems to have been far greater in the case of Vienna and Austria.

In Gödel's day, Vienna still had the only real university in Austria, and this was contained almost entirely in one building. This physical concentration of academic life was indicative of the intellectual life in general. It was a city whose thinkers all seemed to be at least marginally acquainted with one another, influencing each other's thinking across disciplines, so that mathematicians, physicists, historians, philosophers, novelists, poets, musicians, architects, and artists were engaged, in a sense, in the same conversation. The overall topic was the moral and intellectual death and decay of all that had come before, and the need to construct entirely new methodologies, forms, and foundations. It was this sustained theme in so many diverse fields that provoked the emergence of what we have come to call modernity, and even postmodernity: in literature, music, architecture, art, philosophy, psychology, and even, to some extent, science.

Post-1918 Vienna provided a grand tier seat from which to view the rapid disintegration of anachronisms. The Hapsburg Empire, that elaborate variation on the themes of status quo and patriarchy, had imploded with the close of the Great War. Eleven different nationalities—Germans, Ruthenes, Italians, Slovaks, Rumanians, Czechs, Poles, Magyars, Slovenes, Croats, Transylvanians, Saxons, and Serbs—were joined in the ungainly empire; the resultant realm lacked, rather significantly, an accepted name. Its last leader was the long-reigning Franz Joseph, Emperor of Austria since 1848 and King of Hungary since 1867. And the capital city of the nameless realm was Vienna. Though a unifying consciousness eluded the multifarious nationalities of the empire, the "City of Dreams," as Vienna was almost too aptly dubbed, was in fact able to achieve something like that supranational cosmopolitan consciousness entailed by the myth of empire.

Vienna, at its height, had been an imperial capital that had ruled over some 50 million subjects. Now it was the capital of a small and ruined Alpine republic of a little over 6 million citizens, almost all German. (Many of the Czechs who had been living in the city left, which somewhat eased the housing shortage.) But if Vienna was vastly diminished in political terms, its importance as an intellectual capital of the world was unparalleled. The very sense of a spiritual and cultural collapse (whose imminence had been evident to the Viennese thinkers even while the Hapsburg dynasty tottered on) intensified the felt need for the search for new foundations. The collective consciousness of the city's most conscious citizens was suffused with a sort of nervous intensity, the rash of ideas erupting like the symptomatology of diseased genius.

So we find in Vienna not only the birthplace of Zionism in the figure of Theodore Herzl but also of the most extreme manifestation of those ideas that had provoked Zionism as a response, Nazism. It provided the breeding ground for Freud's theory of the unconscious, repression, hysteria, and neurosis; it was where Klimt and Schiele and Kokoschka painted the lushly sensual canvasses of the Secession.[4] Arnold Schönberg and Alban Berg brought forth atonal music and Adolph Loos designed a new sort of architecture, where form would be strictly determined by function, the excessive orna-

4 The Secession was founded in 1897 by artists dissenting from the policies of the Viennese artistic establishment. An exhibition hall for the Secessionists was opened in 1898. Ludwig Wittgenstein's father, the enormously wealthy steel magnate, Karl Wittgenstein, was one of the three "benefactors" whose names are inscribed on the plaque inside the doors. The other two names belong to famous artists of the day: Rudolf von Alt and Theodor Hörmann.

mentation and über-stuffed rooms of the Hapsburg bourgeoisie equated with moral rot.

An influential voice (an acerbic one), sounding throughout Vienna's overlapping cultural circles, belonged to the journalist Karl Kraus, the tireless editor (and primarily the sole writer) of the satirical journal *Die Fackel*, or *The Torch*. Kraus used his journal to clobber every variety of Viennese hypocrisy, whether practiced by the old guard or the avant-garde. (He was, for example, harshly critical of Freud.) Kraus focused much of his fierce crusading attention on language, blasting the deceit that lies curled up in the banalities of respectable forms of speech, the hollowness and insincere sentimentality in works of literature and the empty phrases of journalists. "Speaking and thinking are one," he declared in his book *Die Sprache* (*Language*): The road not only to better theories but to a better society is paved with linguistic precision. Kraus himself was a consummate stylist, skewering his targets in elegant epigrams: "The psychoanalyst picks our dreams as if they were our pockets." "The secret of the demagogue is to appear as dumb as his audience so that these people can believe themselves as smart as he is." "The esthete stands in the same relation to beauty as the pornographer to love, and the politician stands to life."

Kraus's attention to language as the single most important topic in his critique of thought will strike contemporary students of philosophy as so familiar as to seem a truism. Though Kraus was not himself a philosopher, he had a decided impact on Viennese philosophers, and thus on philosophers throughout the world. Ludwig Wittgenstein, in particular, was a regular reader of *Die Fackel*.

Kraus's view that intellectual shoddiness is not only an

offence against truth but also against morality had great cogency among his Viennese contemporaries. A sense of moral urgency underlay their discussion of intellectual and artistic questions, and the exhortatory tone of the Hebrew prophets of old, calling the tribe to repentance, often broke through into discussions of the most dryly abstract sorts of subjects, for example, the conditions for the meaningfulness of propositions. Discredited ideas, solecisms, half-truths, and elaborately phrased nonpropositions carry the fatal taint of moral failure; there is a moral imperative to break with the past and *think clearly.*

The densely swirling culture of Viennese ideas was very much on public display, conducting itself across the round little tables of the city's many cafés. (This, by the way, had something to do with the bleak housing situation in Vienna. The ill-heated and generally inadequate domiciles inclined the Viennese to spend their hours elsewhere.) The literati and other artists favored such places as the Café Museum, the Herrenhof, or the Café Central, where Peter Altenberg, for example, answered so gratifyingly to the popular image of "the poet" in his brightly colored shirt, his wide-striped pants, and his pince-nez dangling poetically from a black ribbon. At another table, Alban Berg and other composers might be found discussing the exhaustion of tonality; or Adolph Loos, discussing the outrages of traditional architecture; or the novelist Franz Werfel, who wrote of the café's "shadowy realm" in his novel *Barbara or Piety.* Another writer, Alfred Polgar, propounded a "Theory of Café Central," explaining that the establishment was a veritable *Weltanschauung,* "a world-view, but one whose essence it was to avoid viewing the world." Its habitués were "for the most part people whose misanthropy was only equaled by their longing for their fellow man, who want to be alone but need company for that."

As for the mathematicians like Gödel, there was the Akazienhof, only a three-minute walk from the university, as well as other places—the Arkadencafé, the Reichsrat, the Schattentor, the white marble tabletops providing a scribbling surface for equations. Not only location, but also the category of people the place attracted and their respective status, as well as the selection of periodicals and newspapers that were offered, influenced where a particular group would gather.

In addition to café society, the intellectual life of Vienna was also organized into various *Kreise*, or circles, more or less formal discussion groups that met on a weekly basis, centered around the leading intellectuals of the city. Many of these circles overlapped. Some were connected with the university, others not. A large number were devoted to discussions of socialism (one, surrounding Max Adler, was Kant-focused), and others were oriented around the various factions within the psychoanalytic movement. A large number of the circles were meant for the discussion of philosophy, not only of Kant, but of such figures as Kierkegaard and Leo Tolstoy, who enjoyed an enormous influence at the time. The philosopher Heinrich Gomperz, in whose class Gödel had become convinced of Platonism, had a discussion group centered on the history of philosophy. The intellectual geometry of Vienna was densely inscribed with circles.

The Vienna Circle

By far the most prominent of these circles was the one that revolved around the philosopher Moritz Schlick, first dubbed, accordingly, the Schlick Kreis, though it came eventually to be known, as an acknowledgment of its preeminence, as the *Der Wiener-Kreis,* the legendary Vienna Circle. It was from this group of thinkers that the influential movement known as "logical positivism" largely disseminated. The reforming edicts of the group reshaped attitudes of scientists, social scientists, psychologists, and humanists, causing them to reformulate the questions of their respective fields; the effects are still with us.

Attendance at the meetings of the Vienna Circle was by invitation only. The philosopher Karl Popper, who went on to eminence and was even then an up-and-coming intellectual force, waited with impatience and in vain for an invitation to join the most important *Kreis* in town.

Kurt Gödel was invited to join while still an undergraduate and was a regular attendant at the weekly sessions between the years 1926 and 1928. Interestingly, 1928 is the year when he turned to mathematical logic, which would of course yield him his famous proof. No wonder he no longer had the time or the inclination for the weekly sessions.

His association with the logical positivists has led to the misconception that he himself was a positivist and that his incompleteness theorems are a consequence of positivist principles. Gödel's incompleteness theorems are still often tallied among positivism's greatest success stories: the revolutionary result of applying its principles to mathematics. So, for example, in the recent *Wittgenstein's Poker,* authors David Edmonds and John Eidinow write:

> The Circle's voice can still be heard in a number of philosophical eponyms. In 1931 Gödel published his theorem that scuppered all attempts to construct a logical founda-

> tion for mathematics. He showed that a formal arithmetical system could not be demonstrated to be consistent from within itself. His fifteen-page article proved that some mathematics could not be proved—that, whatever axioms were accepted in mathematics, there would always be some truths that could not be validated.

Here Gödel's two theorems are more or less correctly stated, though they are merged into one "theorem." But that the Circle's voice can be heard within Gödel's theorems could not be further from the truth. The voice that *Gödel* heard within his theorems was that of Platonism. Any metaphysical position, let alone Platonism, is downright anathema to a logical positivist.

Gödel had become a Platonist in 1925, a year before joining the discussion group. Their anti-metaphysical orientation had no influence on him, and, for their part, they never seemed to suspect—not for a long time at least—that he was not one of them. He apparently gave them little indication. It was not then, and never would be, in his nature to argue face-to-face with those with whom he disagreed. His distaste for engaging in conflict was so extreme as to qualify as an eccentricity, though hardly among his most pronounced. He refused to oppose another person's viewpoint unless he had absolute certainty on his side, unless, that is, he had a *proof*. All his life, he wanted to have his mathematical proofs do all his speaking for him. (Perhaps it is no accident that this man, whose extreme reticence cloaked intense convictions, should have produced the most prolix mathematical results in the history of mathematics.) He was dismayed when others did not catch all that he was trying to say in them. He was dismayed until the end of his life that people still considered his views consistent with those of the Vienna Circle.[5]

What *were* the views of the Vienna Circle? Logical positivism was first and foremost a movement that spoke in the name of the precision and progress associated with the sciences. It sought to appropriate the methodology that had served the sciences so well, to distill the essence of this methodology not only to cleanse science itself of its more mystically vague and metaphysical tendencies—no characterization carried more positivist opprobrium than "metaphysical"—but also similarly to cleanse all intellectual areas. It was a program for intellectual hygiene.

In the Viennese spirit of the time, this group of thinkers from various fields—mathematics, philosophy, the physical and social sciences—were intent on giving the decaying remains of old ideas as hasty a burial as decency required and on resurrecting in their stead a system whose wholesome soundness would derive from the empirical sciences. Logical positivism disseminated out far beyond the little bare room where the group would meet and deeply penetrated the philosophical orientation of philosophers, scientists, and social scientists, many of whom were not even aware that they *had* a philosophical orientation. But the preferred absence of a specifically philosophical orientation was one of the major points emphasized by the logical positivists. It was a philosophical orientation meant to abolish all philosophical orientations, which might strike the reader as paradoxical.

5 Jean Cocteau wrote in 1926 that "The worst tragedy for a poet is to be admired through being misunderstood." For a logician, especially one with Gödel's delicate psychology, the tragedy is perhaps even greater.

Logical positivism is sometimes referred to as "logical empiricism" or "radical empiricism." Traditional empiricism, exemplified by the views of the Scottish philosopher David Hume (1711–1776), had sought to delineate the limits of knowledge. There were, on the one hand, the sort of questions that could be answered through a priori reasoning; these, according to Hume, had no ontological import. They were merely conceptual truths that do not tell us anything about the way the world really is; they merely reflect abstract relationships between concepts. Hume called them "relations of ideas." So the truth that bachelors are unmarried is analogous to the truth that ghosts are the disembodied spirits of the dead and to the truth that fat-free ice cream has no fat. Each is true regardless of whether its subject—bachelors or ghosts or fat-free ice cream—actually exists. On the other hand, there are propositions that reach out beyond the merely conceptual and purport to describe the nature of the world, to say what things exist and what are the properties of, and relations between, things: According to traditional empiricism, any propositions that bear on the nature of the world—Hume called them "matters of fact and existence"—can only be shown to be true or false through the use of empirical means. Some evidence or other is of the essence. The faculty of a priori reason can tell us how our concepts are related to one another, but it cannot tell us what the world beyond our concepts is like. For that sort of knowledge we require some sort of experiential contact with the world.

To use a favorite example, consider the question of the existence of God, defined as a transcendent Being who stands outside space and time, severely limiting the possibilities for experiential contact. (At the least, such experiences would have to occur in time.) Many traditional empiricists had declared the existence of such a trans-empirical God inviolably unknowable, since the cognitive means at our disposal are in principle inadequate for answering the question one way or the other. So remote a God—beyond our experience—may exist, but we'll never know. (Bertrand Russell, when asked what he would say were he to find himself before the pearly gates face-to-face with the Almighty, quipped that his response would be, "Oh Lord, why did you not provide more evidence?")

The logical positivists turned the empiricist theory of knowledge into a theory of meaning. According to the latter, the empirical means that would be relevant to discovering whether a particular proposition is true also provide the very *meaning* of the proposition. The positivist theory of meaning is therefore often called "the verificationist criterion for meaningfulness," and it legislates that the borders of empricial knowability map the borders of meaningfulness. If one cannot, in principle, imagine any possible set of experiences that would count as corroboration for the proposition, then what one has is the mere semblance of a proposition, hollowed out of meaning, what the positivists dubbed a "pseudo-proposition."

By declaring the limits of knowability one and the same with the limits of meaningfulness, the positivists took the problematic aspect of such questions as the existence of God (or of moral values or of abstract entities) up a notch, so that now the unanswerability of certain questions no longer takes the measure of our cognitive inadequacies, but rather signals that the questions ought never have been posed at all. Unknowability is regarded as a sign that a mistake in the use of language has been made. If God (or moral values or universals or numbers) is so defined that no empirical data could possibly be relevant to the question of his (or their) existence, then that question is exposed as ipso facto meaningless: nothing could count as a genuine answer to it. The putative answers—yes, God exists! or no, He does not! are *both* propositional poseurs. Anything that can legitimately be said can be said clearly, the conditions for its meaningfulness one and the same with the conditions for its verifiability (which is not to say that all meaningful propositions are true, of course, but rather that there would be some set of experiences—not forthcoming if the proposition is false—that would establish that the proposition is true). Precision (provided by the verificationist criterion of meaning) becomes the positivist analogue to prayer.

The positivist transformation of the empiricist theory of knowledge into a theory of meaning meant that the single damning word "meaningless" was to be pronounced over the remains of much that had formerly passed for knowledge. Here was the single word with which to accomplish a program of cognitive hygiene such as the world had never seen. The Vienna Circle, which lasted from 1924 to 1930, ending with the tragic murder of Moritz Schlick by a psychotic former student,[6] had an effect that rippled out from Vienna and

6 The student, Johann (or Hans) Nelböck, had already twice been committed to a psychiatric ward for threatening Schlick. He had constructed a delusion in which the influential philosopher was responsible not only for Nelböck's romantic problems but also for his difficulties in finding employment. He shot Schlick on the center staircase of the university's main building (a brass inscription still marks the spot) as the philosopher was hurrying to deliver a lecture to his class. Interestingly, Vienna's Nazis—by 1930 a sizable presence—applauded the psychotic murderer as winning a victory in their own battle for demographic hygiene, with Schlick, a German Protestant, coruscated in the dailies as a godless Jew. Of course, it's

is still actively circulating today, quite often in the introductory "philosophical" chapters of textbooks in science or social science. (The presence in these chapters of such phrases as "a meaningless question because empirically unanswerable" is a dead giveaway. In psychology, for example, the behaviorist school that held dominance for many decades of the twentieth century often asserted that all psychological terms that could not be reduced to the observables of stimulus and response were meaningless.)

Dramatis Personae of the Vienna Circle

Moritz Schlick, if not the most dynamic and innovative of the thinkers of the Circle, was a man whose positivist sincerity and organizational abilities seem to have been instrumental to its success. As philosopher Rudolf Carnap said, "The pleasant atmosphere at the meetings of the Circle was due above all to Schlick's personality, his inexhaustible friendliness, tolerance and modesty." Having trained as a physicist in Germany under the great Max Planck, he had come to Vienna in 1922 to take up the prestigious chair in the Philosophy of the Inductive Sciences at the university, the very chair that had been held both by Ernst Mach and by the towering physicist, Ludwig Boltzmann (for whom Mach's rejection of the molecular hypothesis had constituted both a professional and personal tragedy[7]).

---

true that he was godless, but he wasn't the slightest bit Jewish. In fact, he derived from minor Prussian nobility.

7 Boltzmann had succeeded in deducing the laws of thermodynamics from a statistical analysis of the behavior of a great many molecules. His work was under-appreciated because of the dominant Mach's positivistic

Schlick was sympathetic to the drift of the Viennese über-conversation and his presence at the university soon attracted like-minded thinkers from across many disciplines. At first they gathered in an old Vienna café. But the numbers of those participating gradually grew and, in 1924, Schlick agreed to make the gatherings somewhat more formal, moving the group to a room at the university.

Though all (or almost all) in the Circle held positivist views and everyone (even the clandestine Platonist) had either a connection to or a deep sympathy for the exact sciences, there was a diversity of interests and personalities and opinions among them. There was, for example, Rudolf Carnap, who had been trained as a physicist and mathematician at Jena, where he had been influenced by the logician Gottlob Frege (1848–1925). Carnap was "especially interested in the formal-logical problems and techniques," and would have been a happy man indeed to have seen every question reduced to a straightforwardly technical one—the recalcitrantly irreducible of course declared meaningless. He was said to have had a face, especially in his youth, "that almost seemed to exude sincerity and honesty." His intellectual earnestness impressed his fellow positivists; he worked and learned constantly. When anything came up in conversation that was new to him or that he wanted to follow up, he would produce a little notebook and jot down a few words. His ease in writing soon made him the leading exponent of the Circle's ideas.

Otto Neurath was a social scientist and economist, a great big elephant of a man (he signed his letters with a picture of

---

rejection of the reality of molecules. Boltzmann committed suicide, perhaps partly out of professional despair.

an elephant) with elephantine resources of energy and capacities for enjoying life. Both Carnap (who was an introvert) and Neurath (who was not) had the instinct for political utopianism; and Neurath, in particular, tried to push the Circle in political directions, often making it seem, perhaps unintentionally, that there was more political homogeneity within the Circle than there in fact was. "Schlick especially seemed to resent this since in Vienna, the Circle was named after him, the *Schlick-Kreis*."

Neurath and Carnap felt also that the Circle was intimately connected with other cultural movements, in particular arguing for an affinity between their point of view and the industrial-design-inspired ideology of the Bauhaus. Both were an expression of the *neue Sachlichkeit*, the "fact-mindedness" that received the seal of approval from the sciences. And then in Germany there was the "Berlin Group," centered around the philosopher of science Hans Reichenbach, whose outlook was all but identical with that of the Schlick Circle.

Neurath's sister, the blind, cigar-smoking Olga Neurath, was also an active member of the Circle. She was a mathematician with wide tastes that extended into logic. In her youth she had written three papers, one of which, on the algebra of classes, is described by Clarence I. Lewis in his *Survey of Symbolic Logic* as "among the most important contributions to symbolic logic."

Olga Neurath was married to Hans Hahn, who was also an important member of the Circle. Hahn had been responsible for bringing Schlick from Germany to Vienna. He was a first-rate mathematician, whose name prominently lives on in the useful Hahn-Banach extension theorem in functional analysis. Hahn's mathematical interests were wide, and eventually he became interested in logic. It was he who brought the work in mathematical logic of the German Gottlob Frege and the English Bertrand Russell to the forefront of the Circle's attention. He had an unbounded admiration for Russell and did the Vienna Circle the great service of saving them the difficulty of reading through the monumental three-volume *Principia Mathematica*, explaining it all to them in his seminar of the academic year 1924–25.

Hans Hahn is of particular interest in our story because when Gödel decided to switch his focus from number theory to mathematical logic, Hahn became his dissertation advisor. Though Hahn's specialty was not logic (though he had done some significant work in set theory) his mathematical interests were certainly flexible enough to accommodate Gödel's new interest. Gödel had first come into contact with Hahn in 1925 or 1926, and he told Hao Wang that Hahn had been a first-rate teacher, explaining everything "to the last detail."

Hahn's intellectual interests went far beyond mathematics as well. One of Hahn's extra-mathematical interests was in the empirical evidence for parapsychological phenomena, which was a hot topic in Vienna at this time; a large number of reputed mediums had appeared in the postwar years and eventually a committee was formed, which included Schlick and Hahn and other scientifically oriented thinkers, for the purpose of investigating their claims, the validity of which became a much-vexed bone of contention within the Circle. It was not that Hahn was a believer, but he kept an open mind, which was enough to enrage other members, for example his brother-in-law, Otto Neurath. "Who looks into these matters?" Neurath once demanded with his characteristic vigor, answering his own question in the sociological terms he favored:

"Uncritical, run-down aristocrats and a few supercritical intellectuals such as Hahn. Studies of the kind further the belief in supernatural forces and serve only reactionary groups."

Then there were also two of Schlick's young students, Frederich Waismann and Herbert Feigl, who as "favored students" were invited by Schlick to join the Circle. Hahn, too, would get two of his most talented students, Karl Menger and Kurt Gödel, invited into the select company of Schlick's Circle.

Noch Einmal: Man Is the Measure of All Things

In 1929, when Schlick rejected an offer of a prestigious and lucrative professorship in his native Germany, the other members of the Circle decided to celebrate by publishing, in Schlick's honor, a booklet setting out the tenets and aims of their joint point of view. The result was a sort of positivist manifesto entitled *Wissenschaftliche Weltauffassung: Der Wiener Kreis*, or *The Scientific Worldview: The Vienna Circle.* "Everything," it proclaimed, "is accessible to man. Man is the measure of all things." The ancient Sophist's words were reiterated verbatim, only now given a scientifically minded twist: whatever question is, in principle, not susceptible to measurement, that is, empirical procedures, is no question at all. Since the limits of knowability are congruent with the limits of meaning, no meaningful matter can escape our grasp. We are cognitively complete.

A few years later, Herbert Feigl (who went on to become a prominent philosopher of science in America) co-authored an article in the American *Journal of Philosophy* entitled "Logical Positivism: A New Movement in European Philosophy." The article, Feigl writes, provided "our philosophical movement with its international trade name."

The term "positivism" had long been in circulation, always connoting a pro-scientific attitude used as a standard for meaningfulness. It was first applied to the ideas of Auguste Comte (1798–1857) and Herbert Spencer (1820–1903). The Viennese physicist Ernst Mach (1838–1916) had demanded, in the name of positivism, that all meaningful propositions be reducible to constructs of sense impressions, thus adding much more substance to the positivism of Comte and Spencer. The "Introductory Remarks" chapter of his book *Contributions to the Analysis of Sensation* (1885) has the subtitle "Anti-metaphysical." His positivism had led him to denounce both the reality of atoms and Einstein's relativity. The Schlick positivists acknowledged Mach as one of their guiding lights, although tempering his denunciation of relativity sufficiently to admit Einstein, too, as one of their inspirations (though ignoring the realist interpretation that Einstein conferred on his theory. The positivists, rather, were inspired by Einstein's redefinition of the concept of "the simultaneity of events" in terms of the speed of light).

"Logical" was appended to "positivism," explains Feigl, to emphasize that the Viennese positivists excluded logical propositions (among which they included mathematics) from the otherwise exclusive disjunction: empirical or meaningless.

> The truths of pure mathematics (i.e. not including physical geometry or other branches of the factual sciences) are *a priori* indeed. But they are *a priori* because they are . . . validated on the basis of the very *meaning* of the concepts involved in the propositions of mathematics. The Vienna

> Circle regarded, for example, the identities of arithmetic as necessary truths, based on the definition of the number concepts—and thus analogous to the tautologies of logic (such as "what will be, will be"; "the weather will either change or remain the same"; "you cannot eat your cake and not eat it at the same time").

In other words, the logical positivists believed that mathematics, just like logic, was devoid of any descriptive content. Mathematical propositions, if not quite tautologies, are analogous to them. (It's hard to make out this middle ground, but never mind for now.) Another way to put this point is that mathematics is merely syntactic; its truth derives from the rules of formal systems, which are of three basic sorts: the rules that specify what the symbols of the system are (its "alphabet"); the rules that specify how the symbols can be put together into what are called well-formed formulas, standardly abbreviated "wff," and pronounced "woof"; and the rules of inference that specify which wffs can be derived from which.

To get a sense of what it means to describe mathematics as syntactic (though we will look more closely at the notion in the next chapter), it helps to contrast the view with Platonism. For those who believe that mathematics is syntactic, the phrase "is true" takes on a special meaning when it is applied to a mathematical proposition: A mathematical proposition is true *relative* to the stipulated rules, the syntax, of a formal system. Analogously, for a moral relativist, like Protagoras, "is true," when applied to an ethical statement, is shorthand for "is true relative to *x*," where *x* is a person or, more likely, a conglomerate of ethically agreeing persons. Moral truths are only true relative to the stipulated rules of a society. They are, in the academic terminology du jour, social constructs. Similarly, according to the view under consideration, mathematical truths are formal constructs.

A mathematical Platonist, on the other hand, uses the word "true," even when applied to mathematical statements, in exactly the same way as we normally use the word, not as a shorthand for "relative to *x*," but to represent existing states of affairs. For a Platonist, mathematical truth is the same sort of truth as that prevailing in lesser realms. A proposition *p* is true if and only if *p*. "Santa Claus exists" is true if and only if Santa Claus exists. "Every even number greater than 2 is the sum of two primes," is true if and only if there is no even number greater than 2 that isn't the sum of two primes (even if we can never prove it).[8]

The view, then, of the syntactic nature of mathematics—its lack of any descriptive content—was indicated in the very name "logical positivism." (And Gödel, impassioned Platonist

8 The Prussian mathematician Christian Goldbach (1690–1764) had conjectured that every even number greater than 2 is the sum of two prime numbers (i.e., a number which is only divisible by 1 or by itself). So, for example 4 = 2 + 2, 6 = 3 + 3, 8 = 5 + 3, and so on. Goldbach's conjecture has been confirmed for every even number that has ever been checked; however, no proof has of yet been discovered for the universal conclusion that *every* even number greater than 2 is the sum of two primes. The fact that Goldbach's conjecture remains unproved means (at least according to the Platonist) that lurking out there beyond the point where mathematicians have checked there might be a counterexample: an even number that isn't the sum of two primes. Then again (according to the mathematical Platonist), there may not be a counterexample: every even number may be the sum of two primes, without there being a formal way of proving that this is so. A Platonist asserts that there either is or isn't a counterexample, irrespective of our having a proof one way or the other.

that he was, sat among the positivists and spoke not a word of dissent.)

The scene for the meetings of the Vienna Circle was "a rather dingy room," on the ground floor of the building in the Boltzmanngasse that housed the mathematical and physical institutes of the university. (It is now the meteorological institute.) The room was filled with rows of chairs and long tables, facing a blackboard. When not being used by the positivists, it was a reading room sometimes used for lectures. Those who arrived first at the Thursday evening meetings would shove some tables and chairs away from the blackboard, which most speakers used. The chairs would be arranged informally in a semicircle before the blackboard, and there was a long table in the back used by anyone who wanted to smoke or take notes. People would stand around talking in groups until the signal from Schlick—a sharp clap of his hands. The conversations halted, everyone taking a seat, Schlick at the end of the table closest to the blackboard. He would announce the topic of the paper to be read or the discussion to be pursued, sometimes first reading communiqués from colleagues, and the formal proceedings of the night would begin.

At any one meeting there were usually no more than 20 Viennese members, with foreign visitors sometimes attending. For example, John von Neumann (who, among his other prodigious abilities, managed to inhabit various far-flung points on the globe simultaneously, including Budapest and Princeton) might grace the Circle if he were anywhere in the vicinity; the young Willard van Orman Quine, from America, who went on to dominate Anglo-American analytic philosophy for many decades, from his position at Harvard; Carl Hempel, from Germany, who, among other distinctions, was my first-year advisor when I was a graduate student; the great Polish logician Alfred Tarski (né Teitelbaum), from Poland; and the philosopher Alfred Jules Ayer, from England, all spent time with Schlick's group. Ayer, after spending some months in Vienna and imbibing the doctrines, went back to England and wrote up his imbibition in his highly influential polemic *Language, Truth, and Logic*, thus disseminating the ideas of Vienna's positivists in the English-speaking world:

> We shall maintain that no statement which refers to a "reality" transcending the limits of all possible sense experience can possibly have literal significance; from which it must follow that the labours of those who have striven to describe such reality have all been devoted to the production of nonsense.

By far the most influential figure connected with the Vienna Circle was not even a member of it, and in fact steadfastly refused membership. This was the philosopher Ludwig Wittgenstein. Wittgenstein, at least according to the interpretation that I will propose, plays a significant, if ambiguous, role in the story of Gödel's incompleteness theorems. Wittgenstein's almost mystical influence on the members of the Vienna Circle, the esteemed thinkers among whom the young logician first came to think rigorously about the foundations of mathematics, must have struck a person of Gödel's persuasion as highly dubious. There are still-smoldering remnants of Gödel's resentment of the philosopher to be found in the *Nachlass*, written (though never exposed to the public) many decades after the Vienna Circle had ceased to be, only a few years before the logician's death.

Gödel's and Wittgenstein's views on the foundations of mathematics were, as we will see, at loggerheads, and neither could acknowledge the work of the other without renouncing what was most central in his own view. Each, I believe, was a thorn deep in the other's metamathematics.

Wittgenstein and the Circle

Wittgenstein came from one of the wealthiest and most culturally elite families of Vienna, "the Austrian equivalent of the Krupps, the Carnegies, the Rothschilds, whose lavish palace on Alleegasse had hosted concerts by Brahms and Mahler, Clara Schumann, and the conductor Bruno Walter."[9] He was, in his intensity, preoccupations, ambitions, and conflicts, indelibly stamped by the sensibilities of that intense, preoccupied, ambitious, and conflicted city. While studying aeronautical engineering at the Technische Hochschule in Berlin, he had learned of Russell's paradox, and became interested in the foundations of mathematics.

Russell's famous paradox is of the self-referential variety. The liar's paradox—*this very sentence is false*—is of the same variety. We get into trouble because some linguistic item talks about itself, at least potentially, and by reason of this self-referentiality we end up both asserting that some statement is true and that it is also false, which is logically impossible if anything is.

9 It was a highly musical family. The philosopher's brother, Paul, was a concert pianist who lost his right arm in the First World War. He managed to gain such proficiency with his left hand that he was able to continue his career. Ravel's famous *Concerto for the Left Hand* was written for Paul Wittgenstein.

Russell's paradox concerns the set of all sets that are not members of themselves. Sets are abstract objects that contain members, and some sets can be members of themselves. For example, the set of all abstract objects is a member of itself, since it is an abstract object. Some sets (most) are not members of themselves. For example, the set of all mathematicians is not itself a mathematician—it's an abstract object—and so is not a member of itself. Now we form the concept of the set of all sets that aren't members of themselves and we ask of this set: is it a member of itself? It either is or it isn't, just as the problematic sentence of the liar's paradox either is or isn't true. But if the set of all sets that aren't members of themselves is a member of itself, then it's *not* a member of itself, since it contains *only* sets that aren't members of themselves. And, if it's not a member of itself, then it *is* a member of itself, since it contains *all* the sets that aren't members of themselves. So it's a member of itself if and only if it's not a member of itself. Not good.

Paradoxes have often been found lurking about in the deepest places of thought. Their presence is often the signal (like the canary dying?) that we have managed, sometimes unwittingly, to stumble on a deep and problematic place, a fissure in the foundations. Russell's discovery of his paradox had grievous consequences in the foundations of mathematics, and for one man in particular: Gottlob Frege. Frege had only just finished his monumental two-volume *Grundgesetze der Arithmetic (The Fundamental Laws of Arithmetic)*, which was the first attempt to reduce arithmetic to a formal system of logic. The logic that Frege employed also includes set theory; in other words (to speak in the language of a logician) sets are included as individuals in the universe of discourse, over which the bound variables of the system range. What this basically means is that the system can be interpreted as talking about sets. Numbers are then defined in terms of sets and the arithmetical laws are derived from the axioms and rules of set theory and logic.

Frege's axioms of set theory allowed for the formation of the set of all sets that are not members of themselves, and since this set involves a contradiction—since that set both is and is not a member of itself—there was something fundamentally wrong with Frege's system. Although the system was adequate for the expression of arithmetical truths, it was also inconsistent, which is the very worst thing that a formal system can be. One can prove absolutely anything (and hence effectively nothing) from a contradiction.[10] So an inconsistent system is worthless as a tool of proof.

Russell and his collaborator, Alfred North Whitehead, devised a new formal system for expressing arithmetical truths, accomplished in their *Principia Mathematica* (the very work that appears in the title of Gödel's 1931 paper, setting forth the proof for the first incompleteness theorem). However, to secure consistency, Russell and Whitehead had imposed ad hoc rules governing set formation. Their Theory of Types decrees that there are ascending orders in the universe of discourse—the types of things we interpret the formal theory as speaking

10 Here is why you can prove anything from a contradiction. The rule of inference known as *modus ponens* says that from a conditional proposition of the form *if p, then q* (where *p* and *q* are any random propositions) and from the proposition *p*, you can deduce *q*. Propositions of the form *if p, then q* are false if *p* is true and *q* is false, and true in all other cases. Therefore if *p* is a contradiction, then *if p, then q* will be true for whatever *q* one cares to choose. Therefore, from the assertion of a contradiction *p* anything at all can be made to follow.

about. Basic individuals constitute Type I; sets of individuals, Type II; sets of sets, Type III; sets of sets of sets, Type IV; etc. An item can be a member only of an item of a *higher* type. The question then of whether a set is a member of itself can't even arise. The rules of *Principia Mathematica* bar the formation of such paradox-breeding sets as the set of all sets not members of themselves. Russell and Whitehead called their rules the "Theory of Types," but the problem was that there was no real theory behind the rules at all, as they themselves ruefully acknowledged; there was no explanation at all as to why certain sets were allowable and others not,[11] other than that if one allowed the unallowable very bad things would happen to one's system. Their formal system is consistent by fiat.

Russell's and Whitehead's proffered challenge to logicians to come up with a less ad hoc solution to block the formation of paradoxical sets was what lured Wittgenstein away from aeronautical engineering. This problem that had stumped the great Lord Russell was obviously something worth thinking about. Wittgenstein went to Cambridge, where Russell was the most prominent philosopher on staff, and immediately made himself known to the distinguished philosopher, mathematician, political activist, and aristocrat.[12]

11 Another paradox-breeding set is the set of *all* sets, known as the universal set. The mathematician Georg Cantor (1845–1918) had proved that the power set of a set (formed by taking all the subsets of the set) always has a higher cardinality than the set itself. But then consider the universal set. Clearly, no set could have a larger cardinality than the universal set. However, its power set would . . . a contradiction. This is known as Cantor's paradox, and the rules of set theory must bar the formation of the universal set.

12 He was, at least after the death of his older brother, the third Earl of Russell. His grandfather, Lord John Russell, introduced the Reform Bill of

At first Russell was a bit wary before the strange intensity of the newcomer: "My ferocious German [sic] came and argued at me after my lecture," Russell wrote to his current lover, Ottoline Morrell, the aristocratic wife of the Liberal MP Phillip Morrell. During their affair Russell wrote to her on average three times a day, so there is a lot of useful documentation on which to draw from this period of his life. If only adultery regularly yielded such scholarly riches. "He is armour-plated against all assaults of reasoning. It is really rather a waste of time talking with him." But within a short span of time (while Wittgenstein was still an undergraduate) the "ferocious" convictions of the Austrian had a devastating effect on Russell's confidence in his own logical powers:

> We were both cross from the heat—I showed him a crucial part of what I had been writing. He said it was all wrong, not realizing the difficulties—that he had tried my view and knew it couldn't work. I couldn't understand his objection—in fact he was very inarticulate—but I feel in my bones that he must be right, and that he has seen something I have missed. If I could see it too I shouldn't mind, but, as it is, it is worrying and has rather destroyed the pleasure in my writing—I can only go on with what I see, and yet I feel it is probably all wrong, and that Wittgenstein will think me a dishonest scoundrel for going on with it.

1832, and served as prime minister under Queen Victoria. Bertrand Russell was a political activist, in particular a pacifist. He was jailed twice: once, in 1918, for six months for an allegedly libelous article in a pacifist journal; and again in 1961, at the age of 89, for one week, in connection with his campaign for nuclear disarmament.

The force of Wittgenstein's personality and his reforming attitude toward philosophy, the holy severity of the mission of disabusing his contemporaries of their presumptions (which had a great deal to do with his Viennese sense of the decadent exhaustion of old traditions[13]), transformed Anglo-American philosophy. Like Russell, the Cambridge philosophers and students of philosophy who came to surround Wittgenstein seemed not to have to understand him to know in their "bones that he must be right." His evident brilliance, oracularly (if inarticulately) dispensed against the backdrop of a fierce and formidably austere personality, made for a powerfully convincing display. Wittgenstein quite often gave way to lamentations that his Cambridge colleagues and students did not understand him.

Partly it was the Viennese aspect in his thinking that eluded them. It was not just in his determination to seize hold of a methodology for sweeping away the decay of the old ways and making the entire field new again that he was a Viennese at heart. Wittgenstein's tormentedly dramatic way of pursuing his field, the cult of genius that he propagated, was also highly Viennese. He had read in his youth, and always retained a high respect for, the strange Viennese writer Otto Weininger (1880–1903), "a quintessentially Viennese figure" who had argued that the only way for a man to justify his life (for a woman there is no way) is by acquiring and cultivating

13 Wittgenstein was not apologetic, but perhaps even perversely proud, of how few of the historical philosophical greats he had ever studied. On the other hand, the frontisquotes for his two books "were taken from authors who could hardly have been more typically Viennese—Kürnberger for the *Tractatus,* Nestroy for the *Investigations.*"

genius. Weininger had chosen to shoot himself to death in the very house in which Beethoven, the genius he revered above all others, had died. Wittgenstein himself was suicidal for nine years (his three older brothers committed suicide, also a quintessentially Viennese act), until he came to Cambridge and was pronounced a genius by Russell.

Back in Vienna, Wittgenstein, in absentia, was also producing a profound effect. His first published work, *Tractatus Logico-Philosophicus*, partly written in the trenches of the First World War, had singularly impressed Schlick's group. As stylistically arresting as its creator, this work achieved in its austere elegance a sort of poetry.[14] The traditional tool of the philosopher—the argument—is dispensed with; each assertion is put forth, as Russell once remarked, "as if it were a Czar's ukase." The poet's obscurity of meaning is preserved despite (by means of?) the formal precision of its elaborate numbering system, which hierarchically arranges its assertions: so that, say, proposition 3.411 (*In geometry and logic alike a place is a possibility: something can exist in it*) is an elaboration of proposition 3.41 (*The propositional sign with logical co-ordinates—that is the logical place)* which is an elaboration of 3.4 (*A proposition determines a place in logical space*). The numbering system is borrowed from the mathematician Peano, who had used it in axiomatizing arithmetic, and it is the numbering system that Russell and Whitehead had also employed in *Principia Mathematica.*

14 That it was a sort of poetry was the damning praise of Frege: "The pleasure of reading your book can therefore no longer be aroused by the content which is already known, but only by the peculiar form given to it by the author. The book thereby becomes an artistic rather than a scientific achievement." (From a letter from Frege to Wittgenstein, 16 September 1919, translated in Monk 1990, p. 174)

The Cambridge philosopher G. E. Moore suggested the title, modeled on Spinoza's *Tractatus Theologico-Politicus.* Bertrand Russell wrote the introduction that finally, after much difficulty, secured the author a publisher. Wittgenstein detested the introduction, especially after it was translated into German: "All the refinement of your English style," he wrote Russell, "was, obviously, lost in the translation and what remained was superficiality and misunderstanding." Russell's and Wittgenstein's former intimacy cooled considerably over the following years. "He had the pride of Lucifer," was one of Russell's later summations of Wittgenstein's character.

It was Kurt Reidemeister, a geometer associated with the Circle, who in 1924 or 1925, at Schlick's and Hahn's request, studied the *Tractatus* and suggested that the group read it together.

And so the positivists began a joint study of the *Tractatus,* proposition by proposition, their Thursday-evening meetings now dedicated to Wittgenstein. They read it through not once, but twice, the endeavor taking the better part of a year. (These weekly readings of a portion of the *Tractatus* are reminiscent of the Jewish tradition of weekly readings from the Torah. It happens that 9 out of the 14 original members of the Circle were Jewish by birth, though of course not by conviction—all theistic utterances being regarded as paradigmatically meaningless.)

The Viennese positivists interpreted the cryptic *Tractatus* as offering precisely the new, purifying foundations they sought. Proposition 4.003, for example, could not summarize more perfectly their fundamental conviction:

> Most of the propositions and questions to be found in philosophical works are not false but nonsensical. Consequently

> we cannot give any answer to questions of this kind, but can only point out that they are nonsensical. Most of the propositions and questions of philosophers arise from our failure to understand the logic of our language. . . . And it is not surprising that the deepest problems are in fact not problems at all.

They ascribed to Wittgenstein their own verificationist criterion of meaningfulness, viz. that the meaning of a proposition is identical with the method for verifying it; or, alternatively, that the meaning of a sentence can be reduced to the specification of the experiences that would make the proposition known to be true. They read a vindication of their own positivism in such propositions as 6.53, exhorting one to "say nothing except what can be said, that is, propositions of natural science," and 4.11, "the totality of all true propositions is the whole of natural science (or the whole corpus of the natural sciences)."

They also believed that Wittgenstein had accounted for the truths of mathematics and logic, reducing them to tautologies, devoid of any descriptive content. Proposition 4.461 states that "propositions show what they say: tautologies and contradictions show that they say nothing. A tautology has no truth-conditions, since it is unconditionally true; and a contradiction is true on no condition." There might be terms that refer to items in the world contained in such tautologies as, say, "Socrates is either mortal or he is not." But those referring words are irrelevant to the truth of the tautology. It is the meaning of the purely logical constants—*or* and *not*—that determine the truth of the tautology, and, "4.0132 My fundamental idea is that the 'logical constants' are not representatives; that there can be no representatives of the *logic* of facts." All logic is ultimately tautological: "6.1262 Proof in logic is merely a mechanical expedient to facilitate the recognition of tautologies in complicated cases." Because all logic is tautological, it says nothing: "5.43 But in fact all the propositions of logic say the same thing, to wit nothing."

"6.125 Hence there can *never* be surprises in logic."

(Gödel, of course, was poised to deliver the greatest surprise in the history of logic, one which, in the logician Jaakko Hintikka's words, is "stranger than others by orders of magnitude." So the reader might already suspect, at this early point in the discussion, that Wittgenstein's views on the philosophy of logic, leading as they do to proposition 6.125, would put him starkly at odds with Gödel's result. As we shall see, Wittgenstein never accepted that Gödel had proved what he provably did prove. This, too, might strike the reader as verging on the paradoxical.)

Wittgenstein's discussion in the *Tractatus* of mathematics, as opposed to logic, is brief. Mathematics, he says, is a method of logic (6.2 and again 6.234) and so, presumably, all that he has said of logic applies to mathematics. Mathematics, he says (6.2), also says nothing, has no descriptive content, though because it is expressed in equations it seems to:

> 6.2323 An equation merely marks the point of view from which I consider the two expressions; it marks their equivalence in meaning.
>
> 6.2341 It is the essential characteristic of mathematical method that it employs equations. For it is because of this method that every proposition of mathematics must go without saying.

Mathematical propositions, just like the tautologies of logic, do not represent any facts because they are, in a certain sense, merely grammatical. "6.233 The question whether intuition is needed for the solution of mathematical problems must be given by the answer that in this case language itself provides the necessary intuition." (Proposition 6.233 also puts him starkly at odds with Gödel's result, as we will see.) By language itself, Wittgenstein means syntax, the rules that stipulate that which can be said. Mathematics, like logic, is syntactic. Meanings are irrelevant to the determination of truth, even the meanings of the logical constants and the mathematical "=," for their "meanings" are nothing over and above the grammatical rules that stipulate how we use them:

> 3.33 In logical syntax the meaning of a sign should never play a role. It must be possible to establish logical syntax without mentioning meaning of a sign: only the description of expressions may be presupposed.

Interestingly, from this proposition alone, Wittgenstein claims to demonstrate the fundamental error of the Theory of Types: "3.331 From this observation we turn to Russell's 'theory of types'. It can be seen that Russell must be wrong, because he had to mention the meaning of signs when establishing the rules for them." The next two propositions, 3.332 and 3.333, "dispose of Russell's paradox." Thus Wittgenstein, at least, was satisfied, at least while writing the *Tractatus*, that he had solved the problem that originally lured him into philosophy of logic.

Wittgenstein was later to reject many of the assertions of his *Tractatus*. In fact, the discontinuity in his thinking was judged so radical that he was bifurcated into "early" and "later" Wittgenstein. In place of the early monolithic logic of language, the later Wittgenstein speaks of many different "language-games," each with its own rules. In the early Wittgenstein the interesting nonsense (so to speak), characteristic of philosophy, derives from the violation of the rules that govern the bounds of *all* meaningfulness; in the later Wittgenstein, the interesting nonsense is a result of confusing the rules of one language-game with those of another. (Consistent between the two Wittgensteins is the belief that all philosophical problems arise from confusion about syntax.) The homogeneous view of one language, with one set of rules, from which the positivists took solace, gave way to a postmodern-friendly pluralism of language-games. The later Wittgenstein came to place much more emphasis on the social aspects of rule-following. Rules are embodied in social forms of behavior (also appealing to the postmodern sensibility). Even the law of noncontradiction wasn't to be regarded as an absolute:

> We shall see contradiction in a quite different light if we look at its occurrences and its consequences as it were anthropologically—and when we look at it with a mathematician's exasperation. That is to say, we shall look at it differently, if we try merely to *describe* how the contradiction influences language-games, and if we try to look at it from the point of the mathematical law-giver.

Wittgenstein's attitude toward mathematical logic radically changed. The pluralistic theory of rule-following of the later Wittgenstein was meant to subvert monologicism: that there is but one logic and its name is *Principia Mathematica*. Whereas the early Wittgenstein had labored hard with Russell on problems of logic, the later Wittgenstein came to regard the entire field as a "curse" (while Russell, disheartened by his earlier labors with Wittgenstein—his inability to understand him—withdrew from the field and wrote bestsellers.[15])

Still, even between the early and the later Wittgenstein there is agreement enough on many issues, including fundamental questions in the philosophy of mathematics. Wittgenstein's view of rule-following changed, but he remained committed to the claim that the entire nature of mathematics enfolds from rule-following. All that happens in mathematics is a consequence of rule-following, which is why mathematical "intuitions" are figments of our obfuscation. If we saw clearly what we are doing when we do mathematics we would not resort to these figments.

Both the early and later Wittgenstein are in agreement, then, that there can be no genuine surprises in mathematics. When, therefore, a surprise on the order of Gödel's result arrived, the thing had to be argued away.

Of What We Cannot Speak

Though Wittgenstein may have believed he had summarily disposed of Russell's paradox, the very problem that had drawn him away from aeronautical engineering and into the world of philosophy of logic and language, the entire *Tractatus* constitutes a self-avowed paradox, as the philosopher himself freely admits. According to its own dictates, its

15 His *History of Western Philosophy*, published in 1945, a very comprehensive and readable account of precisely what its title promises, became a longtime bestseller for Simon and Schuster. He turned to such popularizations of philosophy after giving up the more technical work to which he had devoted himself before Wittgenstein entered his life.

very own propositions are meaningless. Wittgenstein forbade talking about a language within the language. The syntactical nature, whether of logic or of mathematics, cannot really, without violating the syntax of the language, be spoken about, but must rather be shown.

> 6.54 My propositions serve as elucidations in the following way: anyone who understands me eventually recognizes them as nonsensical, when he has used them—as steps—to climb up beyond them. (He must, so to speak, throw away the ladder after he had climbed up it.)

(This last metaphor, for which Wittgenstein is famous, was one that Wittgenstein borrowed from the drama critic/philosopher Fritz Mauthner, of whose *Sprachkritik* Wittgenstein tended to be rather critical in the *Tractatus.* 4.0031: "All philosophy is a 'critique of language' [though not in Mauthner's sense"].)

Wittgenstein's attitude toward the inherent contradiction of the *Tractatus* is perhaps more Zen than positivist. He deemed the contradiction unavoidable. Unlike the scientifically minded philosophers who took him as their inspiration, he was paradox-friendly. Paradox did not, for Wittgenstein, signify that something had gone deeply wrong in the processes of reason, setting off an alarm to send the search party out to find the mistaken hidden assumption. His insouciance in the face of paradox was an aspect of his thinking that it was all but impossible for the very un-Zenlike members of the Vienna Circle to understand.[16]

16 Perhaps this is why he found conversation with them so fruitless that on the occasions that he agreed to meet with some of them, he often just turned himself to the wall and read aloud the poetry of Rabindranath

In his autobiography Carnap recalled how the Vienna Circle had struggled with Wittgenstein's dictum concerning the question of whether "it is possible to speak about linguistic expressions."[17] Carnap asked Wittgenstein for elucidation on this point once too often and was summarily banished forever more from Wittgenstein's presence.

Schlick and Waismann were permitted to meet with Wittgenstein in person on a regular basis, Waismann because he was writing a commentary on the *Tractatus*—although Wittgenstein eventually gave up on ever making Waismann understand him and the book was never completed. Waismann was, perhaps, of all the Wittgenstein-enchanted Circle, the positivist who suffered the deepest from philosophical infatuation. He changed his views every time Wittgenstein did, and, like some of the equally impressionable Cambridge students, began to mimic the philosopher's behavioral tics. At each Thursday meeting of the Circle he would update the other members with

---

Tagore, "an Indian poet much in vogue in Vienna at that time, whose poems express a mystical outlook diametrically opposed to that of the members of Schlick's Circle."

17 Carnap was to (prematurely) welcome Gödel's incompleteness theorem as vindicating the Circle's insistence on meaningful metalanguage, and thus as making operable the Circle's positive program for eliminating all metaphysical elements. "By use of Gödel's method," he set about to demonstrate how "even the metalogic of the language could be arithmetized and formulated in the language itself." But it was Gödel himself who soon took the wind out of Carnap's sails, convincing him—or at any rate trying to—that the upshot of the result he had produced by means of his method was entirely at odds with the positivist program itself. "Although Gödel had not persuaded Carnap on this fundamental issue, he *did* move Carnap in a strongly Platonist direction in his definition of analyticity, the capstone of the syntax program."

the breaking news of the philosopher's views, beginning with the disclaimer: "I shall relate to you the latest developments in Wittgenstein's thinking but Wittgenstein rejects all responsibility for my formulations. Please note that." Those members of the Circle whom Wittgenstein refused to meet were thus kept informed, through Schlick and Waissman, of the philosopher's ideas, which were often quoted in their papers. Some Austrian philosophers expressed doubts of the very existence of this "Dr. Wittgenstein" to whom Schlick's group so often made reference. Perhaps he was simply a figment of Schlick's imagination, "a mythological character invented as a figurehead for the Circle."

Just as in Cambridge, Wittgenstein's effect on the logical positivists, particularly Schlick and Waissmann, almost defies explanation. Schlick's wife recalled her husband leaving the house to go to see Wittgenstein for the first time as if he were setting off on a religious pilgrimage. "He returned in an ecstatic state, saying little, and I felt I should not ask questions."

Feigl, in later life, reported, "Schlick adored him and so did Waismann, who, like others of Wittgenstein's disciples, even came to imitate his gestures and manner of speech. Schlick ascribed to Wittgenstein profound philosophical insights that in my opinion were in fact formulated much more clearly in Schlick's own early work."

The note of asperity in Feigl's tone is worthy of comment, since "Feigl had always had an unusual ability to get along with everyone," an ability borne out in his memoirist article, where almost everyone he had ever crossed paths with is described as having only the most amiable personality, the acutest of abilities. Feigl's obliquely expressed distaste for Wittgenstein appears to be traceable back to Feigl's "limitless admiration for Carnap," a systematically precise and conscientious thinker. After banishing Carnap from his presence, Wittgenstein told Feigl, "If he doesn't smell it, I can't help him. He just has got no nose!" When Feigl's admiration for Carnap became clear, he, too, was banished.

Wittgenstein's exasperation with his disciples even in his native Vienna, his insistence that although he might sound like a positivist he decidedly was not one, revolves around the meaning of the closing proposition of his *Tractatus*, numbered simply 7, the severely fulminating (so like a prophet of old): *Wovon man nicht sprechen kann, darüber muss man schweigen,* or: Of what we cannot speak we must remain silent. Schlick's Circle interpreted Wittgenstein as saying in his concluding statement, as well as throughout the book, that the misuse of the conditions of language not only (tautologically) ends in nonsense, but also that outside the bounds of the sayable there is nothing at all; whereas for Wittgenstein there really was "that whereof we cannot speak." The ethical or—what amounts to the same thing for him—the mystical is that whereof we cannot speak. The ethical, the mystical, is both real and inexpressible. He believed that he had explained all that can be said in the *Tractatus*, but as he told one potential publisher (who ultimately passed) what he had *not* said in the *Tractatus*—because it *could* not be said—was more important than what he had said:

> I once wanted to give a few words in the foreword which now actually are not in it, which, however, I'll write to you now because they might be a key for you: I wanted to write that my work consists of two parts: of the one which is here, and of everything I have *not* written. And precisely

> this second part is the important one. For the Ethical is delimited from within, as it were, by my book; and I'm convinced that *strictly* speaking it can ONLY be delimited in this way. In brief, I think: All of that which *many* are *babbling* today, I have defined in my book by remaining silent about it.

He took himself to have demonstrated how little one has actually said after one has finished saying all that can be said.

The question is whether the requisite silence, imposed in proposition 7, hides nothing at all or rather all of the most important things. The positivists certainly interpreted Wittgenstein to be saying the former, which is almost certainly one of the reasons why he dismissed them as not understanding him in the least.

Ironically, the Vienna Circle, united by their core distaste for mystery, were embracing a thinker committed to mystery, at least in so far as questions of ethics, aesthetics, metaphysics, the meaning of life—all the matters they had banished from the realm of reasonable consideration—were concerned. The "unsayable" is, for Wittgenstein, as the "unknowable" was for the traditional empiricists, a measure of our limits. By taking the measure of all that we can say, delimiting it, in his words from within, he is taking the measure of all that we can't say, indicating it without expressing it, since expression is in principle impossible.

Of course, insofar as the sayable was concerned, he truly was propounding a doctrine compatible with positivism, banishing mystery: The meaning of a nontautologous proposition is its method of verification; and so far as mathematical truth goes, Wittgenstein did present a view compatible with the positivists' own, dissolving the seeming mystery of its aprioricity and certainty into the rules of syntax.

Though Wittgenstein raged at the positivists' insistence on fitting him to the procrustean bed of their precision, they mostly responded with adoration, at least in the early days, while Gödel was still attending the Thursday-evening sessions. Olga Taussky-Todd, a mathematician who was Gödel's age and who spent some time in the Vienna Circle, writes, "Wittgenstein was the idol of this group. I can testify to this. An argument could be settled by citing his Tractatus." The visiting A. J. Ayer, who would make good use of his three-months stay in Vienna, wrote back to England, to his friend Isaiah Berlin, in February of 1933, "Wittgenstein is a deity to them all." Bertrand Russell, whom they also respected as an empiricist in good standing, "was merely a forerunner to Christ (Wittgenstein)."

Even the most sober-minded of the positivists, Rudolf Carnap, admits, in his autobiographical notes in the Schilpp volume in his honor, to a measure of near-religious awe:

> When he started to formulate his view on some specific philosophical problem we often felt the internal struggle that occurred to him at that very moment, a struggle by which he tried to penetrate from darkness to light under an intense and painful strain, which was even visible on his most expressive face. When finally, sometimes after a prolonged arduous effort, his answer came forth, his statement stood before us like a newly created piece of art or a divine inspiration. . . . The impression he made on us was as if insight came to him as through a divine inspiration, so that

> we could not help feeling that any sober rational comment or analysis of it would be a profanation.

It was into this oddly Wittgenstein-enchanted Circle (all the odder for being a circle of positivists, the sworn enemies of cognitive bewitchment) that Kurt Gödel was to enter as a student, a reticent observer quietly taking in the opinions around him . . . and drawing his own conclusions.

Gödel in the Vienna Circle: The Silent Dissenter

Regardless of his profound, private disagreements with the positivists, Gödel's association with the Circle led him into the most gregarious few years of his introverted life. He was meeting on a regular basis, not just on alternate Thursday evenings in the bare-bones room where the full Vienna Circle convened but also at late-night sessions in cafés in the garrulous city, men—and the occasional woman—who shared his interest in, if not his intuitions on, foundational issues.

The ever-amiable Feigl reports:

> On the personal side, I should mention that Gödel, together with another student member of the Circle, Marcel Natkin (originally from Lodz, Poland) and myself became close friends. We met frequently for walks through the parks of Vienna, and of course, in cafés had endless discussions about logical, mathematical, and epistemological and philosophy-of-science issues—sometimes deep into the hours of the night.

Karl Menger, in a course on dimension theory he was teaching, had a student by the name of Kurt Gödel, "a slim, unusually quiet young man. I do not recall speaking with him at that time." It was through the regular meetings of the Circle that Menger began the acquaintanceship with Gödel that, though tried, would persist until the end of Gödel's life.

Almost all who were present at those meetings described Gödel in similar terms, as clearly brilliant (though one wonders if this clarity of brilliance emerged in hindsight, after the implications of the announcement of 1930 had fully penetrated) but always quiet, holding his own counsel. Gödel was, according to Feigl, "a very unassuming, diligent worker, but his was clearly the mind of a genius of the very first order." "I never heard Gödel speak in these meetings or participate in the discussions; but he evinced interest by slight motions in the head indicating agreement, skepticism, or disagreement," said Menger, who also reports:

> After one session in which Schlick, Hahn, Neurath and Waismann had talked about language, but in which neither Gödel nor I had spoken a word, I said on the way home: "Today we have once again out-Wittgensteined these Wittgensteinians: we kept silent." "The more I think about language," Gödel replied, "the more it amazes me that people ever understand each other."

What *did* Kurt Gödel think of these "Wittgensteinians?" We know, of course, even though they did not, that he disagreed profoundly with the logical positivists, most specifically on their interpretation of mathematical truth, but far more generally as well. A man whose soul had been blasted by the Platonic vision of truth would not be sympathetic to denunciations of metaphysics. He would not accept a theory of meaning that branded as "meaningless" all descriptive statements that are in principle not empirically verifiable. The *essence* of mathematical Platonism is the claim that mathematics, though not empirical, is nonetheless descriptive. Gödel was a backbencher among the positivists. Though he shared their commitment to precision, as well as their interest in the philosophical relevance of the logical advances of Frege, Russell, and Whitehead, he could not have been more at odds with their metaconvictions. He told Hao Wang many years later (21 November 1971) that the positivists were fundamentally in error in thinking that all meaningful thought could be reduced to sense perceptions:

> Some reductionism is correct, [but one should] reduce to (other) concepts and truths, not to sense perceptions. . . . Platonic ideas are what things are to be reduced to.

In the unsent letter to the sociologist Grandjean, after his opening volley in which he pounced on the sociologist's ingratiating suggestion that his work had been "a facet of the intellectual atmosphere of the early twentieth century," he went on immediately to say:

> It is true that my interest in the foundations of mathematics was aroused by the "Vienna Circle," but the philosophical consequences of my results, as well as the heuristic principles leading to them, are anything but positivistic or empiricistic. See what I say in Hao Wang's recent book "From Mathematics to Philosophy" in the passages cited in the Preface. See also my paper "What is Cantor's Continuum Problem?" in "Philosophy of Mathematics"

> edited by Benacerraf and Putnam in 1964; in particular pp. 262–265 and pp. 270–272.[18]
>
> I was a conceptual and mathematical realist since about 1925. I never held the view that mathematics is syntax of language. Rather this view, understood in any reasonable sense, can be *disproved* by my results.

There we have the matter, stated in an unposted letter. Gödel had become a mathematical realist in 1925, had attended the Vienna Circle's meetings between 1926 and 1928, and by 1928 had begun to work on the proof for the first incompleteness theorem which he interpreted as *disproving* a central tenet of the Vienna Circle, the very tenet that had caused them to append "logical" to the Machian viewpoint of positivism. He had used mathematical logic, beloved of the positivists, to wreak havoc on the positivist antimetaphysical position. Yet here he was, in 1974, still having to explain, in missives that he never mailed, that he was not a positivist, that the intended import of his celebrated theorems had been, in fact, to prove the positivists wrong. The positivists had

18 In the pages referenced Gödel forthrightly sets out his Platonist conviction, using the undecidability of Cantor's continuum hypothesis (that there is no set that is both larger than the set of natural numbers and smaller than the set of real numbers), a mathematical result which he helped to prove, as provocation: "For if the meanings of the primitive terms of set theory . . . are accepted as sound, it follows that the set-theoretical concepts and theorems describe some well-determined reality, in which Cantor's conjecture must be either true or false. Hence its undecidability from the axioms being assumed today can only mean that these axioms do not contain a complete description of that reality. Such a belief is by no means chimerical, since it is possible to point out ways in which the decision of a question, which is undecidable from the usual axioms, might nevertheless be obtained."

endorsed the Sophist's man-measurement of truth. Gödel had sought to vindicate the Sophist's implacable antagonist, Plato.

Gödel, unlike his friend Einstein, did not have a well-developed sense of the ironic, which is, all things considered, a shame.

Gödel and Wittgenstein

It is hard to imagine two more disparate personalities. Wittgenstein and Gödel were both geniuses, both tortured geniuses, in fact. But how they presented that tormented genius to the world could not have been more starkly different.

Wittgenstein had decided views on the nature and duties and privileges of genius. He had once spoken to Russell about Beethoven:

> [A] friend described going to Beethoven's door and hearing him "cursing, howling, and singing" over his new fugue; after a whole hour Beethoven at last came to the door, looking as if he had been fighting the devil, and having eaten nothing for 36 hours because his cook and parlour-maid had been driven away from his rage. That's the sort of man to be.

And Wittgenstein was that sort of man, acting out the high drama of genius, so that Russell, when he was still enthralled, described him to Lady Ottoline as ". . . perhaps the most perfect example I have ever known of genius as traditionally conceived, passionate, profound, intense, and dominating."

He was the sort of genius to attract disciples so fanatical they took to wearing their shirts unbuttoned at the top as he did and aping his tics and mannerisms, such as clapping their hands to their foreheads when struck by a philosophical insight or its lack. They may have disagreed with each other on the *correct* interpretation of Wittgenstein, but agreed that the correct interpretation, if only it could be attained, would almost of necessity be true. (This conviction persists still in significant pockets of Anglo-American philosophy.) Though he more pronounced than argued, still the pronouncements present themselves, singly and in configuration with one another, with the logical austerity craved by rigor-seeking thinkers.

This austerity attached to his person as well, as if the purity of formal logic had been embodied in the man, its standards of absolute truth imposed on human behavior. An anecdote chosen almost at random (there are so many), this one told by Fania Pascal who had known him in Cambridge in the 1930s, bears this out:

> I had my tonsils out and was in the Evelyn Nursing Home feeling sorry for myself. Wittgenstein called. I croaked: I feel just like a dog that has been run over. He was disgusted: "You don't know what a dog that has been run over feels like."

Just as standards of mathematics have made nonmathematicians feel pitifully inadequate in comparison, so the mathematically severe *behavioral* standards Wittgenstein seemed to demand could easily make others feel fraudulent. Wittgenstein seemed, *demonstrably*, the real thing. His was genius turning itself inside out, with all the *Sturm und Drang* of supremely heightened consciousness exposed, and it was a sight before which a thinker, even a logical positivist, might be expected to swoon.

In contrast to Wittgenstein's dramatic display of genius—his Cambridge students recalled that you could *see* the suffering of his thinking—we have Gödel, indicating with slight motions of his head when he agrees, disagrees, is skeptical. His hermetically sealed genius allowing next to nothing of the *Sturm und Drang* of its heightened consciousness to show, Gödel breathed not a word of his fundamental dissent from the beliefs of the Vienna Circle until he had a rigorous mathematical proof to do all his talking for him, until he had mathematical theorems prolix enough to speak out his metaphysical convictions.

It is intriguing to try to imagine the young Gödel, observing his Wittgenstein-bedazzled elders, perhaps more than a little aggrieved or disapproving—not only of the views but also of the very style of the genius so at odds with his own, genius making such a fuss about itself, demanding that others, too, participate in the fuss. One wonders (almost blasphemously) whether there might not have been a bit of human emotion spurring the silent dissenter on to find a *conclusive* refutation, to confront the "divine inspiration" of the philosopher with a higher authority: mathematics.

Such a motivation, however collateral, is conjectural, of course, given the opacity of Gödel's inner life. And we have Gödel's word that Wittgenstein had not influenced his work in mathematical logic at all. In one of the two drafts of the unsent reply to the sociologist Grandjean's questionnaire, in answer to the question, "Are there any influences to which you attribute special significance in the development of your philosophy?", after Gödel had credited Professor Gomperz,[19] he

19 In the alternative set of answers to the questionnaire, he answered the same question, "Are there any influences to which you attribute special significance in the development of your philosophy?" by including the "math[ematical lectures] by Phil. Furtwänger" as well as "phil[osophical] lectures (introductory) of Gomp[erz]."

gratuitously remarked on whom, specifically, had *not* influenced his work: "Wittg[enstein]'s views on the phil[osophy] of math had no inf[luence] on my work nor did the interest of the Vienna Circle in that subj[ect] start with Wittgenst[ein] (but rather went back to Prof. Hans Hahn)."

Of course influence, in a positive sense, is quite different from the sort of murkier incentive I am speculating about. And his adding the gratuitous disclaimer concerning Wittgenstein is telling, especially in so retentive a personality, of at least a retrospective resentment. The influence of the charismatic philosopher on the members of the Vienna Circle may have irked him, amused him (more doubtful), even helped to inspire him in the direction of his proof: we cannot really know. But the older Gödel did leave behind, in the written record, some few scant hints of exasperated pique toward Wittgenstein.

For example, in 1971, mathematician Kenneth Blackwell pointed out to Gödel that there was a passage in Russell's *Autobiography* in which Gödel is mentioned, with various inaccuracies (including his Jewish origins), as well as a rather facile and sarcastic reference to Gödel's Platonism:

> Gödel turned out to be an unadulterated Platonist, and apparently believed that an eternal "not" was laid up in heaven, where virtuous logicians might hope to meet it hereafter.

Gödel drafted a letter, which of course still lies unposted in the *Nachlass*, responding point by point to Russell's inaccuracies on the topic of Gödel [including his alleged Jewishness: "I have to say *first* for the sake of truth that I am not a Jew (even though I don't think this question is of any importance)"] and ending:

> Concerning my "unadulterated" Platonism, it is no more "unadulterated" than Russell's own in 1921, when in the *Introduction* [*to Mathematical Philosophy,* first published in 1919, p. 169] he said "[Logic is concerned with the real world just as truly as zoology, though with its more abstract and general features]." At that time evidently Russell had met the "not" even in this world, but later on under the influence of Wittgenstein he chose to overlook it.

Coming from Gödel, these are pointed words, which is of course why they still languish in a folder in Firestone Library.

Some more decades-smoldering resentment was allowed to escape—this time actually sent off—when his old acquaintance from the Vienna days, Karl Menger, pointed out to him some passages in Wittgenstein's posthumously published *Remarks on the Foundations of Mathematics*, in which Gödel is mentioned. Writes Menger:

> In the early 1970's I began writing a book on my recollections of the Schlick Circle. For the sake of completeness, I looked for ideas about Gödel published by Wittgenstein. A few were in the latter's book *Remarks on the Foundations of Mathematics,* which appeared in 1956. Aside from noncommittal remarks in Part 5, the Appendix I of Part I . . . contains a discussion of the problem—without, however, any adequate appreciation of Gödel's work. In fact, Wittgenstein goes so far astray as to say that the only use of undecidability proofs is for "*logische Kunststücke*" [little logical articifices or conjuring tricks].

After Menger pointed out the passages to Gödel, Gödel responded to Menger:

> As far as my theorems about undecidable propositions are concerned, it is indeed clear from the passage that you cite that Wittgenstein did *not* understand it (or that he pretended not to understand it). He interprets it as a kind of logical paradox, while in fact it is just the opposite, namely a mathematical theorem within an absolutely uncontroversial part of mathematics (finitary number theory or combinatorics). Incidentally, the whole passage you cite seems nonsense to me. See e.g. the "superstitious fear of mathematicians" of contradiction.[20]

These decidedly irked responses to Wittgenstein come *after* Wittgenstein's own reaction to Gödel's famous incompleteness results; and the nature of Wittgenstein's reaction was such as to prompt umbrage even if none had existed before. Wittgenstein never came to accept that Gödel had, through strict mathematics, achieved a result with metamathematical implications. That there could be a mathematical result with metamathematical implications went against Wittgenstein's conception of language, knowledge, philosophy, *everything*. The reticent logician's historically audacious ambitions were, according to the Wittgensteinian point of view, unrealizable in principle. No wonder the anything-but-reticent philosopher would dismiss Gödel's theorems with the belittling

20 Gödel is here citing a parenthetical remark in passage I, 17: "*Die abergläubische Angst und Verehrung der Mathematiker vor dem Widerspruch:* The superstitious fear and awe of mathematicians in the face of contradiction." Gödel also wrote to Abraham Robinson, a young mathematical logician of whom Gödel thought very highly, that Wittgenstein's comments on his proof constitute a "completely trivial and uninteresting misinterpretation" of his results.

phrase "*logische Kunststücke,*" a dismissal that many mathematians have found, to this day, galling in the extreme, as apparently Gödel himself did. (Nary a mathematician I have spoken with has a good word to say about Wittgenstein. One particularly incensed mathematician I know characterized Wittgenstein's famous proposition 7: *Whereof we cannot speak we must remain silent* as "accomplishing the difficult task of being at once portentous and vacuous." The logician Georg Kreisel, who as a student worked with Wittgenstein and later knew Gödel, wrote: "Wittgenstein's views on mathematical logic are not worth much, because he knew very little and what he knew was confined to the Frege-Russell line of goods." Kreisel also decried the transformative influence Wittgenstein had had on students, including himself.)

Yet at a deeper level than even the foundations of mathematics, there was more affinity between the early Wittgenstein's views and Gödel's result than would be apparent from the Vienna Circle's (mis)understanding of Wittgenstein. Wittgenstein really was no positivist, as he insistently protested; and his proposition 7 of the *Tractatus* amounts to a version of his own incompleteness thesis. Of course, to fully appreciate both the substance of the disagreement and affinity between Wittgenstein and Gödel, it is necessary to understand what Gödel actually did. So we will come back to the thorny relationship between Wittgenstein and Gödel later, after the proofs of the incompleteness theorems are presented.

In any case, though Wittgenstein may have loomed large in the circle of men among whom Gödel found himself in his fervently formative years, how seriously the young logician and confirmed Platonist ever took Wittgenstein is, in the end, unknowable. Beyond Wittgenstein towered the figure of the most influential mathematician of the day, David Hilbert, a figure Gödel could not possibly dismiss as mathematically inadequate. Like Wittgenstein's, Hilbert's views on the nature of mathematics could not have been more incompatible with the mathematical result that the young Gödel was soon to spring on an unsuspecting world.

11

Hilbert and the Formalists

A Mathematician's Intuition

We return to the subject of the tantalizing uniqueness of mathematics, pursuing its truth through a priori reason, establishing its conclusions so firmly that no empirical discoveries as to the nature of the world can overturn them.

Since the earliest days of the ancient Greeks, mathematical knowledge has seemed to be on the one hand the least problematic area of human knowledge, in fact the very model toward which all knowledge ought to aspire: certain and unassailable, in short, *proved*. No wonder that epistemological utopians, from Plato onward, have urged that the standards and methods of mathematics ought to be applied, insofar as is possible, to *all* of our attempts to know.

Yet, on the other hand, mathematical knowledge has seemed, to darker-souled epistemologists, highly problematic, its very certainty, which emboldens the utopians, making it suspect in warier eyes. How *can* any knowledge be certain and unassailable, in short: proved? Perhaps, some epistemologists of the darker cast argue, it is because mathematical knowledge is not really knowledge at all; perhaps it is simply a game, played by stipulated rules, telling us nothing about anything. "There is no there, there," Gertrude Stein famously said of her birthplace, Oakland, California. So it is with mathematics, at least according to some.

So the question is: Whence certainty? What is our *source* for mathematical certainty? The bedrock of empirical knowledge consists of sense perceptions: what I am directly given to know—or at least to think—of the external world through my senses of sight and hearing and touch and taste and smell. Sense perception allows us to make contact with what's out there in physical reality. What is the bedrock of mathematical knowledge? Is there something like sense perception in mathematics? Do mathematical *intuitions* constitute this bedrock? Is our faculty for intuition the means for making contact with what's out there in mathematical reality? Or is there just no "there"?

Mathematical proofs must start from somewhere. Often proofs start with conclusions from other proofs and then deduce further conclusions from these. But not everything can be proved, otherwise how can we get off the ground? There must be, in mathematics just as in empirical knowledge, the "given." Given to us through what means? Mathematical intuition is often thought of as the a priori analogue to sensory perception.

Intuitions. They are a tricky business, and not only in mathematics. An intuition is supposed to be something that we just know, in and of itself, not on the basis of knowing something else. (Sometimes, of course, the word is used in a weaker way, conveying the sense of having a vague feeling, lacking in any certainty. But it is in the stronger sense that it functions in epistemological debates.) Obviously intuitions—

or, more precisely, claims to intuitions—vary greatly among people; there are places on the planet right now where people are slaughtering one another because they make claims to fundamentally differing intuitions, patently nonmathematical. All *genuine* intuitions are (tautologously) true (tautologously, because we would not call them "genuine" unless they were true). But not all putative intuitions are genuine intuitions; and how is one to tell when one is in possession of the genuine article? Murky motivations—to believe, for example, propositions that would, if true, conduce to one's own self-importance, notoriously propositions asserting the innate superiority of one's own kind—not only abound but also tend to hide themselves. The resulting beliefs can feel intuitively obvious precisely because we are not prepared to face their real and suspect source in our own personal situations and egos.

You might think that in mathematics—perched on its topmost turret of Reine Vernunft, far from the madding human scene below—murky motives for beliefs are at a minimum. Still, even in mathematics we can get suckered. Accidental features can insinuate themselves into our most pristine mathematical reasoning, presenting us with propositions that seem intuitively obvious when they are not obvious at all—maybe not even true at all.

To illustrate how our "intuitions" can insidiously lead us astray, consider the frequent use of sketches and diagrams to try to make our mathematical abstractions more concrete. These concretizations are almost unavoidable, even to the most mathematically acute minds. For example, David Hilbert, who as we will soon see tried as hard as anyone to impose the strictest rules on mathematics, wrote:

> So the geometrical figures are signs or mnemonic symbols of space intuition and are used as such by all mathematicians. Who does not always use along with the double inequality $a > b > c$ the picture of three points following one another on a straight line as the geometrical picture of the idea "between"? Who does not make use of drawings of segments and rectangles enclosed in one another, when it is required to prove with perfect rigor a difficult theorem on the continuity of functions or the existence of points of condensation? Who could dispense with the figure of the triangle, the circle with its center, or with the cross of three perpendicular axes?

Given that even the most rigorous of mathematicians rely on these aids to pure reason, it can happen that some entirely accidental feature of our sketch gets appealed to in the proof; or that the sketches make it appear that something is just plain obvious when it's not.

Say, for example, you want to prove that the base angles of an isosceles triangle are equal, and by appealing to Sketch 1 below this seems simply obvious, not requiring a proof. Or say you want to prove that all the angles of triangles are acute (less than 90 degrees) and because you have this "truth" in mind you only draw sketches of triangles (Sketch 2) that conform to it. Those are just the only triangles that occur to you.

In other fields of thought as well—for example, and notoriously, ethics—people can have the illusion of "intuitions." In ethics these illusory intuitions can create a great deal of havoc in the real world. Einstein and Gödel were on that quiet little back road in Princeton precisely because it seemed so intuitively obvious to a great many people in Einstein's Germany and Gödel's Austria that the right thing to do was to purify the

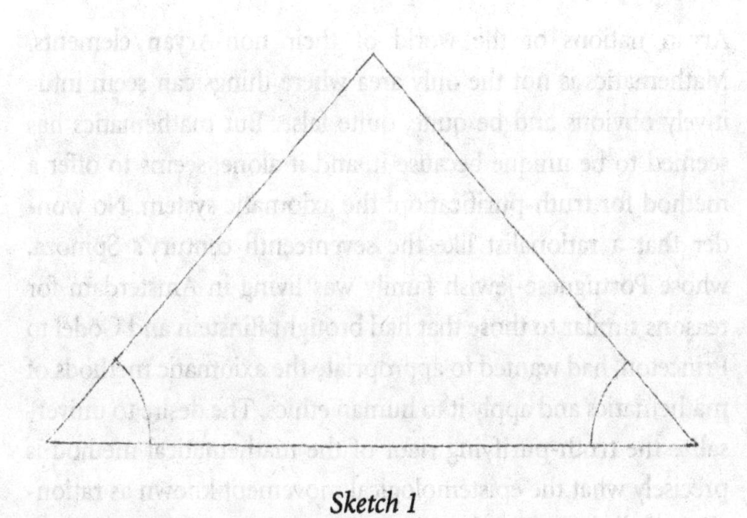

Sketch 1

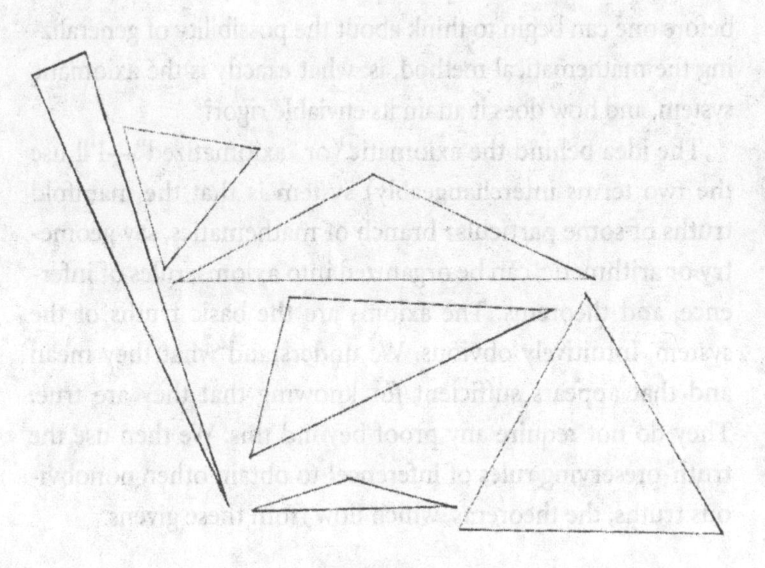

Sketch 2

Aryan nations of the world of their non-Aryan elements. Mathematics is not the only area where things can seem intuitively obvious and be quite, quite false. But mathematics has seemed to be unique because it, and it alone, seems to offer a method for truth-purification: the axiomatic system. No wonder that a rationalist like the seventeenth century's Spinoza, whose Portuguese-Jewish family was living in Amsterdam for reasons similar to those that had brought Einstein and Gödel to Princeton, had wanted to appropriate the axiomatic methods of mathematics and apply it to human ethics. The desire to universalize the truth-purifying rigor of the mathematical method is precisely what the epistemological movement known as rationalism is all about. But the prior question that must be addressed, before one can begin to think about the possibility of generalizing the mathematical method, is: what exactly is the axiomatic system, and how does it attain its enviable rigor?

The idea behind the axiomatic (or "axiomatized"—I'll use the two terms interchangeably) system is that the manifold truths of some particular branch of mathematics, say geometry or arithmetic, can be organized into axioms, rules of inference, and theorems. The axioms are the basic truths of the system, intuitively obvious. We understand what they mean and that appears sufficient for knowing that they are true. They do not require any proof beyond this. We then use the truth-preserving rules of inference[1] to obtain other, nonobvious truths, the theorems, which flow from these givens.

1 These rules of inference are perfect laws of truth-inheritance. The truth that belongs to the ancestors (the axioms) cannot help but be bequeathed to the descendents (the theorems). So, if you know that all *x*'s are *P*'s, and you know that some individual, *i*, is an *x*, then, by the rule of inference known as

For example, consider arithmetic, the simplest branch of all mathematics. Arithmetic concerns the structure of the natural numbers—again, the regular old counting numbers together with 0—and the relationships between them as given by the operations of addition, multiplication, and the successor relation, which takes you from any number n to the number immediately following it in the natural order (i.e., $n + 1$). All other arithmetical operations, such as subtraction and division, can be defined in terms of these three.

In 1889 Giuseppe Peano (1858–1932) reduced arithmetic to five axioms. Here are the first three: 0 is a number. The successor of any number is a number. No two numbers have the same successor. All three appear trivial, which is exactly how we want our axioms. The axioms are so trivial that we can assume that they are true without proving them, with all else following from them, like a huge twisting plant growing out of a simple seed. If we want the whole luxuriant growth to be certain then we want there to be no question *possible* about the truth of the axioms—and this "no question possible" is basically what we mean by "intuitively obvious" or "given" or "trivial" or "self-evident."

The theorems of an axiomatic system, on the other hand, are *only* accepted as true once they are proved, derived from the axioms or derived from other theorems, using truth-preserving rules of inference. Think of it this way, if you care to: Axioms are like the classic first-borns in families: adored sim-

---

"universal instantiation," you know that *i* is a *P*. For example, say you know for certain that all mathematicians do their greatest work before the age of 40, and you also know that Gödel was a mathematician, then you also know that Gödel did his greatest work before the age of 40.

ply for *being*. Theorems are the children that come after, those who have to *prove* themselves worthy to gain acceptance. (First-borns can ignore the analogy. To me, a third-born, the metaphor has a certain appeal.)

So in an axiomatic system (first devised by the ancient Greeks, in particular, Euclid), we begin with a few (the fewer the better) axioms, which are supposed to be intuitively obvious, and then proceed onward to prove whatever follows from these axioms. (The fewer the better, because we want to keep our appeals to intuition to a minimum to maximize certainty.) In place of a libertarian policy of "let's-just-depend-on-the-good-intentions-(intuitions)-of-citizens-(mathematicians)-to-do-the-right-thing," the axiomatic system imposes some strict governmental controls. In place of random appeals to intuitions, there is to be general consensus on what is directly given, the bedrock, with everything else subjected to systematic rule-regulation. You can think of axiomatization as sort of "big government mathematics." The motive behind the axiomatic system is to maximize certainty by minimizing appeals to intuitions, restricting them to the few ineliminable axioms. But the latter are crucial because, after all, we do have to start somewhere.

For much of the history of Western thought, at least since the time of Euclid, the axiomatized system was generally deemed to represent mathematics—and thus knowledge itself—in its most perfect form. Gottlob Frege, who further simplified Peano's axiomatic system for arithmetic by deriving Peano's five axioms from a single axiom, said: "In mathematics we must always strive after a system that is complete in itself." It is this system-building that accounts, Frege said, for the unique certainty of mathematics and "no science can be so enveloped in obscurity as mathematics, if it fails to construct a system."

The drive for limiting our intuitions went even further. The aim became to eliminate intuitions altogether. This aim is what brings us, at long last, to the notion of a *formal system*. A formal system is an axiomatic system divested of all appeals to intuition.

Why take the drastic step of intuition-divestment? Well, intuitions, as we said before, are a tricky business. Though *genuine* intuitions are true, how can we tell when we are in possession of the genuine article? Maybe we can't. Maybe the feel, the urgent cogency that compels belief, is exactly the same whether the intuition is for real or is not. Then what good is the appeal to intuition? So, all things being equal, it would seem a good thing to rid ourselves of these appeals, especially when pursuing the "severest of all disciplines."

In fact, all things were not equal, and the inequality was such as to give added impetus to the drive to ruthlessly eliminate intuitions from mathematics. The nineteenth century gave us mathematical developments that subverted our confidence in those intuitively obvious givens of our axiomatic systems. (Firstborns can go terribly wrong.) The most dramatic of these undermining events was the discovery of non-Euclidean geometry. This unanticipated mathematical development demonstrated that one of the givens of Euclid's geometry, the notorious parallels postulate, is not so axiomatic after all; in fact it is possible to construct self-consistent geometries in which it isn't even true![2] Then set theory, too, delivered us some nasty news about

2 The fifth of Euclid's five postulates was the notorious parallels postulate, which states that through any point outside a line, only a single line can

our putative intuitions. The givens of set theory, again so intuitively obvious, lead to the formation of such paradox-infected sets as the set of all sets that are not members of themselves.

Clearly the bedrock consisting of our mathematical intuitions was not much of a bedrock after all. If it was possible to

---

be drawn parallel to the original line. Euclid himself wasn't all that happy with this last postulate, sensing how different it was from the others, with its covert reference to infinity, and he had avoided using it in his derivations whenever he could. Why does the parallels postulate invoke infinity? Two lines are parallel if and only if they'll never intersect. But if you take a finite region of space then you can draw more than one line through a point that will be parallel (i.e., won't intersect) the line. So the parallels postulate makes implicit reference to infinity, and we are always rightly suspicious of our intuitions about infinity. Euclid's suspicion about this one element of his system (his masterpiece was entitled *The Elements*) was duplicated down through the ages, with various mathematicians attempting to convert the problematic axiom into a theorem by deriving it from the other four axioms. Then, in the nineteenth century, mathematicians changed their tactics, attempting to show that the fifth postulate followed from the other four indirectly: by taking the four and the negation of the fifth and seeing whether a contradiction could be derived. Instead of a contradiction, an entirely new and self-consistent geometry was derived! Three mathematicians independently derived non-Euclidean geometry: the incomparable Carl Friedrich Gauss (1777–1855), known as "the prince of mathematicians"; Nicolai Ivanovich Lobachevsky (1792–1856); and the young János Bolyai (1802–1860), who on stumbling on this new mathematical world in 1823 wrote to his father, Farkas Bolyai, himself a mathematician and a friend of Gauss, "I have discovered things so wonderful that I was astounded. . . . Out of nothing I have created a strange new world." When Gauss was shown the results, he wrote, "I regard this young geometer Bolyai as a genius of the first order," but he had to inform the young genius that he was not the first to derive such a strange new world. He himself had done so, but had suppressed the results because he felt them to be too controversial.

purge our axiomatic systems of appeals to intuitions altogether then that was the way to go.

The elimination of intuitions is accomplished by draining the axiomatic system of all meanings, except those that can be defined in terms of the stipulated rules of the system. The rules, in terms of which everything else is defined, make no claim to being anything other than stipulated. They make no pretense of being descriptive of some objective reality, of independent objects like numbers and sets. A formal system is precisely what we are left with after this meaning-drainage. This deprivation constitutes further "governmental controls," the most stringent that mathematicians could think of, so that no appeals to intuition come sneaking in. You can think of it as the Communist takeover of mathematics, abolishing private property (meanings), everything taken over by public rules.

A formal system, then, is an axiomatic system—with its primitive givens (the axioms), its rules of inference, and its proved theorems—except that instead of being constructed of meaningful symbols—such as terms referring to the number 0 or to the successor function—it is constructed entirely of meaningless signs, marks on paper whose only significance is defined in terms of *the relations of each to one another as set forth by the rules.* While pre-purged axiomatic systems were understood as being about, say, numbers (arithmetic) or sets (set theory) or space (geometry), a formal system is an axiomatic system that is not, in itself, *about* anything. We don't have to appeal to our intuitions about numbers or sets or space in laying down the givens of the formal system. A formal system is constituted of stipulated rules: that specify the symbols ("alphabet") of the system; that tell us how we may combine the symbols with one another to produce grammatical configurations (wffs); and that tell us how we may proceed to deduce wffs from other wffs (the rules of inference).

The formalization of axiomatic systems was meant to offer the highest standard of certainty so that we don't have to depend on our intuitions as to what is mathematically obvious and what is not. It was meant to obviate our reliance on mathematical intuition altogether, to turn our mathematical activity into processes so completely determined by clearly specified rules as to be purely *mechanical*, requiring no imagination or ingenuity, not even a grasp as to what the symbols mean. To follow the rules of a formal system—and a formal system consists of nothing but rules—is to engage in a *combinatorial* activity that, consisting purely of *recursive* functions (roughly speaking, functions that tell you how to arrive at a result by taking the result of another recursive function, or of a really simple basic function[3]), could be programmed into a computer, that is, that is *computable*. This activity amounts to figuring things out by using an *algorithm*,[4] a sequence of operations that tells you what to do at each step, depending on what the outcome of the previous step was.

As the previous paragraph aimed to indicate, a whole family of interrelated mathematical concepts makes its appearance with the move toward formalization. The concepts of a mechanical or an effective procedure, of recursive and computable functions, of combinatorial processes and of an algo-

3 The very fruitful mathematical concept of a recursive function was first defined by Gödel in his proof of the first incompleteness theorem.

4 The word comes from the name of the ninth-century Persian mathematician Abu Ja'far Mohammad ibn Mûsâ al-Khowâsizm, who wrote an important mathematical book in about AD 825 that was called *Kitab al jabr w'al-muqabala*. We also derive our word "algebra" from the title of his book.

rithm: this family of concepts all mean pretty much the same thing, revolving around the idea of rules that are applied to the results of prior applications of rules, all with no regard to any meanings except for what can be captured in the rules themselves.

Intuition can get no dangerous foothold in a formal system. Intuitions tell us what to think about actual things—about space, about numbers, about sets. We don't have intuitions about made-up, meaningless symbols and the rigid rules we have set down for manipulating these. We don't need them. Everything our a priori reason needs to do in a formal system is specified by the rules, which is why the idea of a formal system is so closely connected with the idea of the computer, with what it is that computers can do and how it is that they do it. This is why the concept of the computable is part of the tangle of concepts surrounding the notion of formal systems.

Formal systems, even if their implications are sufficiently convoluted so that real mathematical cunning is required to get to them, have a transparency that precludes intuition. Intuition is (reputedly) a way of trying to circumnavigate the essential opacity of actual things, a way of making contact with them, and it's a way that had proved itself to be eminently unreliable in mathematics just as elsewhere. A mathematics done formally is a mathematics purged of any "given" truths—those claiming an unquestionable source in the "true nature of things," in and of themselves.

If it could be shown that logically consistent formal systems are adequate for proving all the truths of mathematics, then we would have successfully banished intuitions from mathematics. (The proviso of "logically consistent" is of course necessary, since from inconsistent systems one can prove anything at all.) We would have shown, too, that mathematics should not be considered as inherently about anything. By banishing intuitions we would be dissolving away the putative objects of mathematical descriptions. We would be showing mathematics not to be descriptive at all.

The assertion of the possibility and desirability of banishing intuitions by showing formal systems to be entirely adequate to the business of mathematics is the metamathematical view known as *formalism*.

In formalism's retelling, mathematics becomes chess raised to a higher order of intricacy. There is, we can all agree, no objective chess reality that the system of chess captures. The stipulated rules constitute the whole truth of chess. Similarly, according to formalism, the stipulated rules constitute the whole truth of mathematics. We win in mathematics by proving theorems—that is, by showing some uninterpreted string of symbols to follow from other uninterpreted strings of symbols, using the agreed-upon rules of inference. There is no external truth against which mathematics has to measure itself.

Gödel's first incompleteness theorem states the incompleteness of any *formal system* rich enough to express arithmetic. So Gödel's conclusion, you might suspect, has something to say about the feasibility (or lack thereof) of eliminating all intuitions from mathematics. The most straightforward way of understanding intuitions is that they are given to us by the nature of things; again intuition is seen as the a priori analogue to sense perception, a direct form of apprehension. So Gödel's conclusion, in having something to say about the feasibility (or lack thereof) of eliminating all appeals to intuitions from mathematics might also have a thing or two to say about the actual existence of mathematical objects, like numbers and sets. In other words, the adequacy of formal systems—their consistency and completeness—is linked with the question of the ultimate eliminability of intuitions, which is linked with the question of the ultimate eliminability of a mathematical reality, which is the defining question of mathematical realism, or Platonism. It is because of these linkages that Gödel's conclusions about the limits of formal systems have so *much* to say. This is how they got to be the most verbose theorems in the history of mathematics and how they were understood, at least by their author, to assert the metamathematical position to which he had given his heart and soul. The young student had found a proof for a theorem, the first incompleteness theorem, that had the rigor of mathematics and the reach of philosophy.

Mathematical verbosity, as opposed to verbosity of any other sort, could not have better suited the personal eccentricities of Kurt Gödel, a man who had so much to say on the nature of mathematical truth and knowledge and certainty, but wanted to be able to say it using only the rigorous methodology of mathematics. With a proof in hand, he would not have to involve himself in the sorts of combative human conversations he regarded with distaste, maybe even with horror. There never was a man, I'll wager, who combined so much conviction with so little inclination to argue his convictions by the normal means given to us, viz. human speech.

The irony of course is that while his theorems were accepted as of paramount importance, others did not always hear what he was attempting to say in them. They heard—and continue to hear—the voice of the Vienna Circle or of existentialism or postmodernism or of any other of the various fashionable outlooks of the twentieth century. They heard everything except what Gödel was trying to say.

Math Goes Formal

The leading advocate of formalism was David Hilbert, who was the most important mathematician of his day. "Mathematics," wrote Hilbert, "is a game played according to certain simple rules with meaningless marks on paper." His proposal to formalize one branch of mathematics after the other, starting with the most basic branch of all, arithmetic, came to be called the Hilbert program. The successful completion of the Hilbert program would offer significant vindication of formalism, explaining the sui generis aprioricity of mathematics as derivative from the stipulation of rules.

Mathematicians, according to formalism, are not in the business of discovering descriptive truths, whether of the real world of things in physical space or the trans-empirical world of numbers and sets. They were never really meant to discover, for example, how many lines parallel to a given line can run through a given point in space that isn't on that line. They are simply in the business of manipulating the mechanical rules of self-enclosed formal systems that are complex enough to test the deductive skills of mathematicians.

In spirit Hilbert's formalism was close to the antimystery attitudes of the Vienna Circle, and so the logical positivists very naturally embraced it, as Feigl tell us:

> With the formalists (*e.g.* Hilbert) we would consider mathematical proofs as procedures that start with a given set of

> sign combinations (premises, postulates) and according to rules of inference (transformation rules) lead to the derivation of a conclusion.

Formalism confirms, at least in the sphere of mathematics, what the positivists had declared in their manifesto: viz. that man is the measure of all things. We create our formal systems and all of mathematics follows.

For centuries, utopian epistemologists, like Descartes and Leibniz, had been inspired by the unique certainty and aprioricity of mathematics and had harbored dreams of extending those very features throughout the cognitive realm, obviating recourse to empirical evidence, which can give one, at best, mere probability. The special features of mathematical truth had led otherwise sober men, going back all the way to Plato, to near-mystical celebrations of its otherworldly reach. But now, with formalism, the nature of math's aprioricity and certainty were claimed to derive from nothing more mystical than the stipulated mechanical rules of meaningless formal systems, replicable by the soon-to-be-invented electronic gadgetry. With the success of Hilbert's program, the foundations of mathematics would at last be laid cleared, the murkiness that had encouraged rationalist giddiness dispelled.

In 1899, David Hilbert published *Grundlagen der Geometrie,* or *Foundations of Geometry,* said to have been the most influential work in geometry since Euclid's. Its importance reached far beyond geometry. He had shown that geometry could be captured in a formal system—conditional, that is, upon arithmetic's being formalized, since geometry, like all branches of mathematics, presumes the truths of arithmetic. (In this sense, arithmetic is the most basic of all mathematical systems.)

In 1900, a year after he had published his *Grundlagen der Geometrie,* Hilbert gave the keynote address at the Second International Congress of Mathematicians, held that year in Paris. The date, inaugurating a new century, was important. In the talk, called "Mathematical Problems," Hilbert delegated himself the task of determining what the next century would bring by way of mathematical achievement. He laid out 23 problems that he considered the most important to solve.

In his introduction, which has the distinct tone of a pre-game pep talk, Hilbert urges on his "team" of mathematicians with the assurance that no matter how difficult a particular problem may seem, victory is inevitable:

> This conviction of the solvability of every mathematical problem is a powerful incentive to the worker. We hear within us the perpetual call: There is the problem. Seek its solution. You can find it by pure reason, for in mathematics there is no *ignorabimus.*

Hilbert describes this as a conviction "which every mathematician shares, but which no one has as yet supported by a proof." Then he stated the 10 problems.

It is of some interest that Gödel, even though a logician—*merely a logician,* according to the mathematical biases that persisted far into his own day (and dim echoes of which still sound today)—contributed enormously to the first two problems, as well as to the tenth, that Hilbert, *the* establishment figure in mathematics, determined as the most outstanding problems to be solved.

The first of these is the problem revolving around Cantor's continuum hypothesis. What is Cantor's continuum hypothesis? The great nineteenth-century mathematician Georg Cantor had proved that (roughly speaking) there are more real numbers than there are natural numbers, even though there are an infinite number of both. Cantor showed, by means of an elegant argument called the "diagonal argument," that in the infinite pairing of one natural number with one real number, every natural number will be paired with a real number, but some real number or other will forever be left out. The set of real numbers is thus of a higher ordinality (roughly speaking, there are more of them) than the set of natural numbers. Cantor hypothesized that there is no infinite set that intervenes between the set of natural numbers and the set of real numbers; that is, there is no set that has a higher ordinality than the natural numbers and a lower ordinality than the real numbers. This is Cantor's continuum hypothesis and Hilbert's first problem was to prove Cantor's continuum hypothesis true. Gödel was to contribute to the solution to this problem, though not in a way that Hilbert welcomed. Gödel, together with Paul Cohen, proved that the continuum hypothesis can be proved neither true nor false within current set theory. In other words, the status of the continuum hypothesis is what Hilbert claimed there could not be: an *ignorabimus*—a claim that can neither be confirmed nor discredited, a claim about which we remain ignorant.

But it is Hilbert's second problem which is of particular concern to us. Here, too, Gödel's solution of this problem could not have been less welcome to Hilbert.

Hilbert's Second Problem: The Consistency of Arithmetic (The Most Important Proof That Wasn't to Be)

Hilbert's second problem was to prove the consistency of the axioms of arithmetic. For a system to be consistent means that it does not yield any logical contradictions.

The urgency of the problem of consistency was a direct result of the veer toward formalism. When axioms were understood to be asserting true statements about actual objects, there was not as pressing a worry about inconsistency. When you said something was an axiom, you took it for granted that it was true in the most naïve sense of "true": that is, that it described some state of affairs. This meant that neither the axiom, nor any theorem derived from the axiom, could possibly logically contradict any other axiom or theorem, because they were all true statements in the old-fashioned, prosaic sense of "true": descriptive of actual things. True and precise descriptions about reality cannot be logically incompatible with one another.

Think of it this way: If I am truthfully describing my apartment—that it is located in New York City, that it has (alas) only one bathroom—I don't need to stop and worry whether some of these statements contradict each other, whether, for example, I will be able to derive that I live in suburban New Jersey and have four bathrooms. If all my statements are both unambiguous and truly descriptive, then they won't contradict each other, since the objective truth of the matter underpins them all.

But in a formal system, with axioms drained of meaning, and truth amounting to nothing beyond provability, it cannot be taken for granted that the axioms will not yield logically incompatible theorems. Our formal systems are constituted by stipulated rules. Who is to say that we—mere humans, after all—constructed these systems consistently, that the rules will not have contradictory implications? This is the downside of taking man as the mathematical measure of all things, with no independent reality to ensure the ultimate consistency of our axioms.

Of course, if my axioms do lead to theorems that are logically incompatible, then my system is worthless, worse than the most speculative chain of barely probabilistic conjectures or of metaphysical artifices claiming to scope out the ends of Being and Ideal Grace. Anything at all can be deduced within an inconsistent system, since from a contradiction any proposition can be derived. You might say that on a strictly formalist interpretation, inconsistency loses its sting. What's so awful if, through the manipulation of meaningless wffs, we arrive at contradictory meaningless wffs? The game is ruined, of course, since it just isn't interesting to try to derive theorems if everything's derivable; but it's not as if inner contradictions are disastrously demonstrating that our systems can't be *true*—the elimination of the extra-systematic truth being the whole point of formalism. You might also say that the exigency with which Hilbert urged that mathematics be proved consistent showed that he wasn't really, deep down, a formalist after all. In any case, following the formalist agenda, if mathematics is to be successfully purged of intuitions in the service of certitude, then formal proofs of the *consistency* of the purged systems are a pressing necessity.

The highest priority of all was to prove the consistency of arithmetic. Other systems of mathematics, for example geometry, had been shown to be consistent *provided arithmetic, the most basic of all mathematical systems, is consistent.* These sorts of proofs, which establish the consistency of one system as a consequence of the consistency of another system, are called relative proofs of consistency. All of these relative proofs of consistency were related to the consistency of arithmetic, which thus became the next step, the one that would provide the linchpin for the Hilbert program. The proof of arithmetic's consistency could not be relative, as the other consistency proofs were; it had to be an "absolute proof."

The tone of Hilbert's 1900 talk to the mathematical troops is extremely optimistic; he felt pretty sure that an absolute proof of the consistency of arithmetic was imminent. But in a series of talks which Hilbert gave through the 1920s, his buoyancy modulated into something more guarded. His change in mood was brought about by the paradoxes of set theory, including most conspicuously Russell's paradox of 1902, which brought the set of all sets that are not members of themselves to the horrified attention of mathematicians, adding insult to the injury of Cantor's paradox, involving the impossible universal set, the set of all sets.[5] Hilbert regarded "the situation with respect to the paradoxes" with dismay:

> Admittedly, the present state of affairs where we run up against the paradoxes is intolerable. Just think, the definitions and deductive methods which everyone learns, teaches, and uses in mathematics, lead to absurdities! If mathematical thinking is defective, where are we to find truth and certitude?

5 See note 11 in chapter I.

But he still continues to express his confidence that there is "a completely satisfactory way of escaping the paradoxes without committing treason" against the spirit of mathematics.

"Treason" against the spirit of mathematics would consist of such acts as expelling the very notion of the infinite from mathematics in the way that mathematicians such as the Dutch Luitzen Brouwer were advocating. Brouwer was a leading exponent of the intuitionist school of mathematics, yet another metamathematical outlook.[6] The intuitionists were the most fundamentally opposed to Platonism of all the non-Platonist schools. Even formalism can be interpreted in such a way so that it doesn't preclude a basically Platonic approach to mathematics, just so long as Platonically sanctioned intuitions do not in any way enter into mathematical practice.

6 The intuitionists were the most severe of all when it came to the question of acceptable methods of proof. Mathematical proofs were to be limited, according to the intuitionist, to "constructive proofs," i.e., those that employed concrete operations on finite or "potentially" (but not actually) infinite structures. Reference to completed infinite structures were forbidden, as were indirect proofs making reference to the law of the excluded middle. Using the severe strictures of intuitionist mathematics a great deal of accepted mathematics, for example parts of classical analysis and even classical logic, would be deemed no longer acceptable. Brouwer himself renounced much of the former work he had done before becoming an intuitionist convert. "Intuitionism," by the way, might seem like a misleading name, considering the way we have been speaking of intuitions up until now, as just the sort of appeals to objective mathematical truth that formalists and intuitionists meant to eliminate. The intuitionists claimed that their finitary constructions were actually mental constructions, and in fact, the only sort of mathematical mental constructions that we, being finite, could actually perform. So they were claiming that their strictures on mathematical proofs actually corresponded to human psychology.

Hilbert himself often sounds like a Platonist, albeit one who is very strict with himself.

Hilbert certainly wanted to avoid the extreme limitations that the intuitionists would place on mathematical practice in order to avoid the possibility of paradoxes. The approach that Hilbert had in mind, when he spoke of "completely clarifying *the nature of the infinite*," lay within the reach of finitary formal systems.[7] In other words, the way out of the morass visited upon mathematics by the paradoxes of set theory lay, according to Hilbert, in the purification process of formalization—that is, in solving the problems of what Hilbert dubbed "proof theory."

Chief among these was the problem that Hilbert placed second in his list in the 1900 talk, that of finding a finitary proof of the consistency of the axioms, first of arithmetic, then of progressively stronger axiomatic systems. It would all begin, then, with the "single but absolutely necessary" step of proving the formal system of arithmetic consistent. The work first of Gottlob Frege, and then, after the discovery of Russell's paradox, of Russell and Whitehead in *Principia Mathematica*, had prepared the ground for this ultimate sanctifying proof. The formal system laid out in *Principia Mathematica* was sufficient for expressing all the truths of arithmetic; it was also, presumably, consistent. The ad hoc rules of the Theory of Types barred the formation of inconsistency-breeding sets

7 In speaking of formal systems so far in this book, we've been speaking of *finitary* formal systems only; i.e., formal systems with a finite or denumerable (or countable) alphabet of symbols, wffs of finite length, and rules of inference involving only finitely many premises. (Logicians also work with formal systems with uncountable alphabets, with infinitely long wffs, and with proofs having infinitely many premises.)

like the set of all sets not members of themselves. Still a formal proof of consistency was a necessity. Then, if it could be shown that such a formal system was both complete, allowing us to prove all arithmetical truths, as well as consistent, the linchpin of the Hilbert program would have been secured, the crisis posed by the paradoxes overcome.

Enter Gödel.

III

The Proof of Incompleteness

Gödel in Königsberg

What the logician Jaakko Hintikka had called Gödel's *Sternstunde*, his shining hour, occurred 7 October 1930. The scene was the third and last day of a conference in Königsberg on "Epistemology of the Exact Sciences," which had been organized by the *Gesellschaft für empirische Philosophie*, the Organization for Empirical Philosophy, an association that overlapped both with the Vienna Circle and the Society for Scientific Philosophy, a Berlin discussion group, whose leading light was Hans Reichenbach, a philosopher of physics. The aims and activities of the Berlin group were similar to those of the Vienna Circle and close ties existed between the two groups from the beginning. Some of the most influential mathematicians, logicians, and mathematical philosophers had been invited to deliver papers at the conference. Gödel, who had only just completed his Ph.D. dissertation, was not one of the big fish. He was scheduled, together with other small fry, to give a 20-minute talk on the second day of the conference.

On the first day there were four speakers, each talking on behalf of a distinctive metamathematical position. The meta-concern addressed was, as is almost always the case when discussing metamathematics, the tantalizing aprioricity and certainty of mathematics. How have we been allowed membership in the most selective cognitive country club going?

Rudolf Carnap traveled by train from Vienna with Gödel to deliver a paper entitled "The Main Ideas of Logicism," presenting the view that mathematical truths are ultimately reducible to the tautologies of logic. Arend Heyting, a Dutch mathematician, spoke on "The Intuitionist Foundations of Mathematics," urging the banishing of all but strictly constructivist proofs, not making reference to any notions that are not strictly finite or at least denumerable. (The result would be to eliminate a great deal of beautiful mathematics.) David Hilbert, the leading spokesperson for formalism, did not travel from Göttingen, but his formalist point of view had an eminently worthy spokesman in John von Neumann. There was also a paper entitled "The Nature of Mathematics: Wittgenstein's Standpoint" by the long-suffering Wittgensteinian epigone, Frederich Waismann.

Logicism, intuitionism, formalism, Wittgenstein: there was no representative of Platonism to argue that point of view on the first day in Königsberg. All the views represented there that day were committed to the claim that the notion of mathematical truth was reducible to provability; the disagreements between them were on the conditions of provability.

It had been because Waismann was preparing for his talk in Königsberg that Wittgenstein had been meeting with him and

Schlick, in Schlick's house, on a regular basis in the summer of 1930. The central points of Waismann's lecture, according to Wittgenstein's biographer Ray Monk, were: "... the application of the Verification Principle to mathematics to form the basic rule: 'The meaning of a mathematical concept is the mode of its use and the sense of a mathematical proposition is the method of its verification.'"

Monk goes on to say: "In any event, Waismann's lecture, and all other contributions to the conference, were overshadowed by the announcement there of Gödel's famous Incompleteness Proof."

This is not, in point of fact, how it went at the conference in Königsberg. It is understandable that Wittgenstein's biographer would assume that the announcement of Gödel's "famous Incompleteness Proof" would have caused a sensation among the participants at the conference, who had heard talks on that first day incompatible with such a result as Gödel's, each talk presuming that the concept of mathematical truth is, one way or another, reducible to provability. But Gödel's "announcement" went almost unheard.

It is true that Waismann's talk did not hold its own among the other three, but this is because, as Menger and others report, the other participants agreed that Wittgenstein's views were not yet ripe enough for debate. Waismann was not shoved into the shadows because Gödel took the limelight. In fact, Gödel's announcement, delivered during the summarizing session on the third and last day of the conference, was so understated and casual—so thoroughly undramatic—that it hardly qualified as an announcement, and no one present, with one exception, paid it any mind at all.

Gödel's First Great Overlooked Moment: A Triviality Not So Trivial

The 20-minute talk that Gödel had delivered on the second day of the conference had also attracted little attention. It was basically a précis of the work he had done the year before for his Ph.D. dissertation, work not on incompleteness but rather on completeness. What Gödel had done was to prove the completeness of what is called the "predicate calculus" or sometimes "first-order logic" or again "quantificational logic." Never mind the ugly names, causing poetic souls to duck for cover. Let's rechristen the relevant system of formal logic "limpid logic." Gödel proved that limpid logic is complete. Its axioms and rules of inference allow one to prove all logically true, or tautologous, propositions within it. But what is this notion of logically true or tautologous?

Limpid logic's symbolism allows one to represent propositions so that they are stripped down to their naked logical form. It provides a way of symbolizing the logical form of propositions and displaying the logical connections between them. It has symbols for such words as *not*, *and*, *or*, *if . . . then . . .*, *if and only if* as well as such "quantificational" concepts as *all*, *none*, and *some*. Words such as these are the logically relevant ones. It's the meanings of these terms, as defined by the rules of the system, that determine the logical form of propositions. Different sentences can share the same logical form and, from the point of view of limpid logic, these sentences are essentially the same, since they are logically the same (thus continuing the move toward logical generality, which is one and the same with the development of the science of logic fathered by Aristotle.

So, for example, consider the sentences "all married men are married," "all beautiful babies are beautiful," "all valid arguments are valid." All of these sentences are spun out from the more general proposition that if something has two predicates P and Q, say, being both a baby and being beautiful, then it has one of those predicates, say, being beautiful.

Limpid logic categorizes whole hosts of sentences in terms of their shared logical form, stripping away all the meanings of specific predicates and subjects to get down to the naked logic. So far as the nonlogical terms go, these refer either to individuals—whether to specific ones or to any of them—and to predicates and relations between them. To refer to any of the individuals we use variables like x and y, and to refer to specific individuals we use constants, like a or b. Properties are designated by predicate constants like P and Q, and then there are relational terms like R. The easiest thing we can say in limpid logic is that some individual has some predicate: $P(a)$. We read this as: P of a. A slightly more complicated statement is that some individual bears a particular relation to another: $R(a\ b)$. Then we might want to say something like: there is some individual or other that has a particular property. This is symbolized as $\exists xP(x)$, and is read: there exists an x such that P of x. Or we might want to say that all individuals have P. In limpid logic this is symbolized as $(x)P(x)$, and is read: given any x, P of x. A logically true proposition, or a tautology, is one that is true no matter *what* meanings we substitute for the nonlogical terms. (Since "logically true" thus makes reference to meanings—something is logically true if it's true no matter what meanings we assign to its nonlogical terms—it's a semantic, rather than syntactic, notion.)

So, for example, suppose we want to say that if something

has two specific predicates then it has one of those predicates. We would symbolize this:

$$(x)\ (P(x) \text{ and } Q\ (x)) \rightarrow Q(x)^1$$

This is read: given any *x*, if *x* has the property *P* and *x* also has the property *Q*, then *x* has the property *Q*. This is logically true and will generate a whole heap of true propositions, formed by substituting in particular meanings for *P* and *Q*.

Here's another logically true proposition from limpid logic:

$$(x)\ (y)\ ((x = y) \rightarrow (P(x) \leftrightarrow P(y)))$$

This means: for all *x*, for all *y*, if *x* equals *y*, then *x* has the property *P* if and only if *y* has the property *P*. Of course it doesn't take too much thought to see that this has to be the case, that is, that if two things are really not two but rather identical, then all the properties of the one are the properties of the other. (What we really have in the case of identity, is one thing being designated, or picked out, in two ways.)

From this last formally true proposition of limpid logic we can get out such verities as: If Gödel is the author of the incompleteness theorems, then Gödel is a Platonist if and only if the author of the incompleteness theorems is a Platonist. If Professor Moriarty is the mastermind behind all the crime in London then Professor Moriarty is a mathemati-

1 Notice that parentheses are used for punctuation in limpid logic. ~(*p* & *q*), or "it is not the case that both *p* and *q*," is an altogether different proposition from ~*p* & *q*, or "not *p* is true and also *q*." So "it's not the case that the president is both good-natured and stupid" is not the same assertion as "the president isn't good-natured and he's stupid."

cian if and only if the mastermind behind all the crime in London is a mathematician. If the moon is the goddess Diana, then the moon is made out of green cheese if and only if the goddess Diana is made out of green cheese. Obviously, you can fill in anything at all for your predicate *P* and generate a trivially true statement, because $(x)\ (y)\ (x = y) \rightarrow (P(x) \leftrightarrow P(y))$ is logically true. It's a tautology, its truth a function of the meanings of the logical terms that compose it.

Gödel's completeness theorem, the result he presented to the conference of logical luminaries, proved that all such logically true propositions are provable within the formal system of limpid logic. Another way of stating Gödel's completeness result is that in limpid logic syntactic and semantic truth are equivalent: the truths that follow from the rules of the system (the syntactic truths) yield all the logically true propositions expressible within the system. Limpid logic is, then, not only consistent (its consistency had already been proved) but also complete. (Inconsistent systems are of course complete, because we can prove anything at all in them. They're *over*complete. It's of consistent systems that the question is posed: are they complete? Do their formal syntactic rules allow one to prove everything one would like to be able to prove? Do they allow one to prove all the truths that are expressible within the system?)

Completeness is exactly what one would like from one's formal system of logic, and it was one of the problems for which Hilbert had demanded a solution. It was reassuring to have a proof, but since the conclusion had never really been in doubt, Gödel's Ph.D. result hardly seemed exciting. The young man had taken the trouble to prove what everyone already took for granted.

In hindsight, we can see that what Gödel had proven in his

dissertation was far more interesting—cause for far more concern among formalists and fellow travelers—than it had first appeared. Gödel had proven the expected result—completeness—but the difficulty of proving it—the substantive proof, of many steps, it required—should have struck people as unexpected, even alarming. In showing how complicated it was to actually prove the completeness of limpid logic Gödel was certainly creating room for the possibility that *other* consistent formal systems, those, for example, enriched by the axioms of arithmetic, might not be complete. The nontriviality of the proof of completeness for limpid logic must have forcefully presented the possibility to Platonist Gödel that there were propositions that were *arithmetically* true but not provable within a formal system of arithmetic.[2]

2 Gödel was many years later to write Wang that his completeness proof for the predicate calculus, i.e., his dissertation problem, had also been guided by his Platonist convictions: "The completeness theorem, mathematically, is indeed an almost trivial consequence of Skolem 1922 ["Some Remarks on Axiomatized Set Theory"]. However, the fact is that, at that time, nobody (including Skolem himself) drew this conclusion. . . . This blindness is indeed surprising. But I think the explanation is not hard to find. It lies in a widespread lack, at that time, of the required epistemological attitude toward metamathematics and toward non-finitary reasoning. Non-finitary reasoning in mathematics was widely considered to be meaningful only to the extent to which it can be 'interpreted' or 'justified' in terms of finitary metamathematics. (Note that this, for the most part, has turned out to be impossible in consequence of my results and subsequent work.) According to this idea metamathematics is *the* meaningful part of mathematics, through which the mathematical symbols (meaningless in themselves) acquire some substitute of meaning, namely rules of use. Of course, the essence of this viewpoint is a rejection of all kinds of abstract or infinite objects, of which the prima facie meanings of mathematical symbols are instances."

The Quietest Explosion: Gödel Announces His Result

Gödel gave no indication of the revolution he was hiding up his sleeve until the last day of the conference, which had been reserved for general discussion of the papers of the two preceding days. He waited until quite late in the general discussion and then he mentioned, in a single immaculately worded sentence, that it was possible that there might be true, though unprovable, arithmetical propositions, and moreover that he had proved that there are:

> One can (assuming the [formal] consistency of classical mathematics) even give examples of propositions (and indeed of such a type as Goldbach and Fermat[3]) which are

3 "Goldbach and Fermat" refer, respectively to "Goldbach's conjecture" and "Fermat's last theorem." Goldbach, as was mentioned in footnote 8 in chapter I, had conjectured that all even numbers greater than 2 are the sum of two primes. The French mathematician Pierre de Fermat (1601–1665) had written in a margin of a book, found after his death, that he had "discovered a truly marvelous demonstration of the proposition" that there are no integers x, y, z, n, with $n > 2$, such that $x^n + y^n = z^n$, "which this margin is too narrow to contain." At the time of Gödel's announcement, neither Goldbach's conjecture nor Fermat's last theorem had been proved either true or false, though generations of mathematicians had ardently tried. In 1991, Andrew Wiles of Princeton University succeeded in demonstrating, in a complicated proof that required the results of many other mathematicians and took more than 150 pages, Fermat's last theorem. Goldbach's conjecture has still neither been proved nor disproved. (Disproof would be easy enough: finding an even number that isn't the sum of two primes.) The possibility that Gödel was asserting in his Sternstunde was that such propositions as these two may, in fact, be true but unprovable within formalized number theory. What his famous proof does, of course, is to pro-

really contextually [materially] true but unprovable in the formal system of classical mathematics.

That was it. The proof that was to become the "famous Incompleteness Proof" had apparently been accomplished the year before, when Gödel was 23, and it was to be submitted in 1932 as his *Habilitationsschrift*, the last stage in the prolonged process of becoming an Austrian or German *Dozent*. It is one of the most astounding pieces of mathematical reasoning ever produced, astounding both in the simplicity of its main strategy and in the complexity of its details, the painstaking translating of metamathematics into mathematics by way of what has come to be called Gödel numbering. It is a thoroughly ordered blending of several layers of "voices," both mathematical and metamathematical, counterpoint merging into harmonic chords never before heard. Music does seem to provide a particularly apt metaphor, which is why Ernest Nagel and James R. Newman in their classic explicatory work, *Gödel's Proof*, described the proof as an "amazing intellectual symphony."

It must have been an extraordinarily exhilarating experience to have produced such mathematical music, especially since it is mathematics that sings, at least in the ear of its composer, of his beloved Platonism. But Gödel had not let a single note of his symphonic proof escape until this muted moment in Königsberg. Such a noiseless, inexpressive exterior enclosing such a swelling mathematical noise. Then, at long last, he pronounced one tersely precise sentence, dropped in medias

duce such a proposition, one that can be seen to be true even as it is proved to be unprovable.

res on the last day of a conference, in the middle of the rehashing of the previous days' pronouncements. Gödel brought it out with no fanfare, played it barely *pianissimo*.

The idiosyncratic "announcement" is congruent with the logician's personality. The concise statement that composed his "shining hour," lasting maybe 30 seconds tops, is meticulously crafted, a miniature masterpiece. It says what it needs to say, and not a word more. He must have prepared it "to the last detail" (to echo the encomium he bestowed on Hahn's lectures) and given careful thought, too, as to the precise moment when he would launch it into the discussion: toward the end of the three days, as the conclusive refutation of all the metapositions heretofore argued. The man who indicated with slight movements of his head when he agreed, disagreed, or was skeptical, must have thought that the vast significance of his remark would emerge in stunning relief for the audience at Königsberg.

Delivery, even on the most rigorous subjects, can make quite a difference to the reception of one's ideas. Self-importance helps; a heavy, elaborately carved frame can make the sketchiest artwork seem important. We have no first-hand accounts of the manner of Gödel's presentation that October day in 1930; of the sort of expressive framing he gave to what was the mathematical analogue to a painting representing the nature of beauty itself. But we know enough about the emphatically anti-charismatic Gödel, with his aversion to external drama and his absolute faith in logical implication, to be able to imagine how it went. The somber and uninflected statement of the crux of the matter, with no rhetorical flourishes, no hyped-up context to help his listeners grasp the importance of what was being said. No *Sturm und Drang,* only zipped-up genius emitting an austere sentence that implied the existence of a proof of unprecedented nature and scope.

"The more I think about language," he had remarked to Menger, walking home after an evening with the "Wittgensteinians," "the more it amazes me that people ever understand each other." Such pessimism about the possibilities for communication—even at this early age, before the decades that brought so much celebratory misunderstanding of his work—must certainly have stoked his desire to find a strict mathematical proof to say all that he had to say on the nature of mathematical truth and knowledge. Now he had such a proof and he was announcing its result, or at least the first of the two theorems to follow from it. Had he anticipated that dumbfounded disbelief would follow the dropping of his bombshell, and then a violent volley of targeted questions? Had he prepared himself conscientiously for all the demands for further elucidation that would be expected quite naturally to follow, much as Wittgenstein's biographer had imagined the scene?

Gödel was always to be disappointed by the abilities of others to draw the implications he had scrupulously prepared for them, and his experience at Königsberg must have been a magnificent disappointment, for the response was a resounding silence. The immaculately composed sentence was delivered . . . and the discussion proceeded as if it hadn't been. The edited transcript of the discussion that day was published in the journal *Erkenntnis* (edited by Carnap and Reichenbach, and the main organ for the dissemination of the views of both the Vienna Circle and Reichenbach's Berlin group) and it does not include any discussion of Gödel's remark at all. No mention of Gödel made it into the account of the meeting written up by Hans Reichenbach either.

Even taking into account Gödel's anti-charismatic mode of being in the world, shouldn't his remark have provoked a ripple of disturbance, an "*Excuse me, Herr Gödel, but I somehow thought you just said that you'd proved the existence of unprovable arithmetical truths. Of course, you couldn't have been saying that because, besides flying in the face of all of our views on the nature of mathematical truth, that sounds like a contradiction in terms. How could you prove that there are arithmetical propositions that are both unprovable and true? Wouldn't that proof, in showing them to be true, constitute a proof of them, thus contradicting your claim that the proof proves them unprovable? Logician that you are, you couldn't be asserting a blatant contradiction like that. So what did you really say?*"

Gödel's dissertation advisor, Hans Hahn, was present at Königsberg. In fact, he chaired the last-day discussion at the conference. Had Gödel shared nothing of his incompleteness proof with his advisor? We don't know for sure either way. Hao Wang writes that Gödel completed his dissertation, the completeness proof for predicate logic (i.e., limpid logic), without showing it to Hahn. That result was being prepared for publication at the time of the conference (and, presumably, Hahn had read it by then). In his introductory remarks to the dissertation, which for some unknown reason were deleted from the published version, Gödel had raised the possibility of the incompleteness of arithmetic, though he gave no indication that he had proved it. Hahn must have read that remark and not taken it seriously. Maybe it was he who advised the young author to delete it. Maybe he didn't want his student to go out on a fragile limb with no proof to support him, having assumed, from all that Gödel had told him (or, more relevantly *hadn't*), that there *was* no supporting proof.

Another person at the conference who might have been expected to react to Gödel's remark was Rudolf Carnap. Carnap had had more time than the others to digest Gödel's news, since Gödel had confided his result to Carnap several weeks before. On 26 August 1930, according to Carnap's *Aufzeichnungen* in his *Nachlass,* he had met with Gödel, Feigl, and Waismann in the Café Reichsrat in Vienna to discuss their travel plans to Königsberg. After settling the practicalities, the discussion turned, in Carnap's words, to "Gödel's *Entdeckung: Unvollständigkeit des Systems der PM; Schwierigkeit des Widerspruchsfreiheitbeweises*: Gödel's discovery. The incompleteness of the system of *Principia Mathematica*. The difficulty of proving consistency." (Note that he says "difficulty" here, not yet impossibility. Gödel didn't fully prove his second incompleteness theorem until after the conference.) Then again, three days later, Carnap records that the same four met at the same café and that before Feigl and Waismann arrived "*erzählt mir Gödel von seinen Entdeckungen*—Gödel told me about his discoveries." Still, at the conference Carnap had pushed his old line, that consistency was the sole criterion for judging the adequacy of formal theories of mathematics, with the question of completeness not even raised. How, given Gödel's *Entdeckungen,* could he not have questioned his former thinking?

The answer seems to be that Carnap had not understood the nature of Gödel's discovery at all. The idea that the criteria for semantic truth could be separated from the criteria for provability was so unthinkable from a positivist point of view that the substance of the theorem simply could not penetrate.

This delay in grasping the significance of what Gödel was trying to tell him is borne out from another entry in Carnap's journal, dated 7 February 1931, after Gödel's famous paper had already been published. "*Gödel hier. Über seine Arbeit, ich sage, dass sie doch schwer verständig ist.* Gödel was here. About his work, I say that it's very difficult to understand."

The thundering silence greeting Gödel's announcement seems, in retrospect, a classic example of the sort of insensibility that Thomas Kuhn discusses in *The Structure of Scientific Revolutions:*

> In science . . . novelty emerges only with difficulty, manifested by resistance, against a background provided by expectation. Initially only the anticipated and usual are experienced even under circumstances where anomaly is later to be observed.

Von Neumann Takes the Hint

There happened to have been one person present at Königsberg who picked up on the anomalous remark of the young logician, and that was John von Neumann. His appreciation of Gödel's terse remark is all the more impressive if we consider that von Neumann's own views were entirely in keeping with Hilbert's—he had been the designated spokesman in Königsberg for formalism—and that he harbored the sort of strongly positivist bent that would make Gödel's reference to semantic truth, independent of a formal system, perhaps seem dangerously metaphysical. Nonetheless he buttonholed Gödel after the discussion ended for the day and pumped him for details. Gödel must have told him enough about how he had arrived at his conclusion for von Neumann to take what he heard seriously. He went back to Princeton, to the Institute for Advanced

Study, and continued to ponder the astounding pronouncement he'd heard in Königsberg.

Some time in the course of his pondering, von Neumann happened on a remarkable corollary to what Gödel had told him. Von Neumann had seen from what Gödel had told him that Gödel's proof was conditional: what it says is that *if* a formal system S of arithmetic is consistent, then it's possible to construct a proposition, call it G, that's true but unprovable in that system. So if S is consistent, G is both true and unprovable. Trivially, then, if S is consistent then G is true. Von Neumann had also understood from what Gödel told him that this proof can itself be carried out in a system of arithmetic. (This is the trick that's accomplished by Gödel numbering.) So if the consistency of S could be proved in S, then G would have been proved in S—since it follows from the consistency of S that G is true. But this contradicts that G is unprovable. The only way out of the contradiction is to deny that S can be formally proved to be consistent within the system of arithmetic. So from Gödel's result another impossibility follows: it is impossible to formally prove the consistency of a system of arithmetic within that system of arithmetic.

Von Neumann got in touch with Gödel, informing him of this corollary, and Gödel politely told von Neumann that the older man had indeed drawn the correct conclusion, one which Gödel had already rigorously proved. (One can imagine Gödel's slight crooked grin in imparting this information to the intellectual titan, von Neumann.) This corollary is known as Gödel's second incompleteness theorem, and though it's merely a consequence of the first, it's the one that first received attention, with von Neumann talking it up at the Institute. Hilbert's program had provided the context for perceiving the significance of Gödel's second incompleteness theorem. Gödel had proved that Hilbert's second problem could not be solved: there would never be a finitary formal proof of the consistency of the axioms of arithmetic within the system of arithmetic. There would never be the proof that was to serve as the linchpin for Hilbert's program. The consequences for formalism were stark and devastating.

It's clear from the semantic point of view—that is, when we think about what the strictly formal system of arithmetic means—that arithmetic is consistent, since it has a model in the natural numbers. We give a model of a formal system, also called an interpretation, by specifying a universe of discourse, also called a domain of individuals, over which the variables range, and by interpreting what the predicate and relational terms mean, and stipulating which members of the domain the individual constants refer to. When the formal system of arithmetic is endowed with the usual meanings, involving the natural numbers and their properties, its axioms are seen to be clearly true and thus must be consistent since true statements can't have false consequences. It isn't that the consistency of arithmetic was really in doubt—that is, if one really does believe in numbers. The question concerns how consistency can be proved—an important question from any metamathematical point of view, an *urgent* question for all but the Platonist.

The consistency of a system amounts to the proposition that using the rules of the system no contradiction can be derived. This proposition is itself combinatoric; it concerns simple rules of symbol manipulation—rules that determine which string of symbols follow from which strings of symbols. This combinatoric proposition is, precisely because it is combinatoric, equivalent to something arithmetical. Thus it can itself be formulated within the system—and so the question naturally is whether it can be proved within the system, and the answer is that it can't be. The syntactic features of formal systems—which were meant to obviate intuitions, those breeders of paradox—can't capture all the truths about the system, including the truth of its own consistency. Gödel had proven that consistency, the very thing whose proof was supposed to secure the foundations of Hilbert's program (which had been meant to preclude the formation of paradoxes in mathematics) transcended the grasp of that program.

The possibility of paradox, meant to be forever eliminated by Hilbert's program, reasserted itself. And one of the strangest things about the odd and beautiful proof that subverted Hilbert's defense against paradox was the way in which paradox itself was incorporated into the very structure of the proof.

The First Incompleteness Theorem: The Overall Strategy

The twenty-odd pages of Gödel's famous proof are densely compact. There are 46 preliminary definitions. There are also preliminary theorems that must be proved before the main event can take place: the construction of an arithmetical proposition that is both true and unprovable within the formal system under consideration. The lines of reasoning in the proof are highly compressed, composed of a hierarchy of interconnected levels of discourse, the blended voices of the symphony.[4]

4 There are, first of all, the statements within the formal system S under consideration (call them S-sentences); when you interpret these S-sentences under the natural interpretation (i.e., as being about the natural numbers)

Though the details of the proof are difficult, the overall strategy is—happily!—almost simplicity itself. Simple but strange, as one would expect of a proof that draws so close to the edge of self-contradiction, *proving* that there are *true* arithmetical propositions that are not *provable*. One of the strangest things of all about the proof is that it co-opts the very structure of self-referential paradoxes, those abominations to reason, and reshapes that structure to its own ends. The overall strategy of the proof—the delightfully accessible part of the proof—can be grasped in the context of the oldest known paradox of all, the liar's paradox.[5] We can convey the

---

they turn into arithmetical statements (call them A-sentences). The S-sentences each get numbered with the so-called Gödel numberings. Then there are the metamathematical statements about the S-sentences and about the formal system (call them M-sentences). M-sentences, which describe the purely formal relationships between the elements of the formal system, are combinatoric statements, so in a sense you might almost think of them as already mathematical statements. But it takes the ingenious encoding system of Gödel numbering to transform the M-sentences into A-sentences, so that in talking *about* the formal system of arithmetic (M-sentences) you are also making arithmetical statements (A-sentences).

5 In Gödel's famous paper of 1931, in which the proof is first set forth, he mentions the liar's paradox and Richard's paradox, offering them to us as heuristic grips for hoisting ourselves up into the strange country of his proof. We are already acquainted with the liar's paradox. Richard's paradox was the creation of French mathematician Jules Richard. It's rather a complicated one to state, requiring, much like Gödel's proof itself, a certain sort of mapping. One orders the properties of the natural numbers and assigns a number to each of the properties. The number assigned to a property may or may not actually have that property. So, say, 22 corresponds to the property "being an even number." Then 22 itself has the property to which it corresponds in the Richardian ordering. Now define this property: "not having the property assigned to itself in the Richardian ordering." Call this property "being

gist of the proof in a breezy, easy way, which we will now proceed to do. Then in the next few sections we will concentrate on filling in a few more of the details, indicating how the hard work gets done.

By tradition, the liar's paradox is attributed to the Cretan Epimenides, who reputedly said something implying: *All Cretans are liars.* This sentence, in itself, isn't paradoxical, except insofar as it suggests that what Epimenides was saying was something like this:

This very sentence is false.

Now that sentence, as we've already seen, is true if and only if it's false—not a good situation, logically speaking. Gödel's strategy involves considering an analogue to that paradoxical sentence, viz. the proposition:

This very statement is not provable within this system.

Let's call this sentence *G*. G, unlike its analogue, isn't paradoxical, though it is, like all self-referential propositions, somewhat strange. (Even the nonparadoxical self-referential *This very statement is true* is mystifyingly strange. What's it saying? Where's its content?)

Richardian." The paradox-generating question is: is the number that corresponds to being Richardian itself Richardian?

All other formulations of his proof—for example, those by Alan Turing and G.J. Chaitin—have incorporated features of paradoxes, though different paradoxes from Gödel's—into their own versions of the proof. These paradoxes, though different from one another, are all of the self-referential variety. The affinity between the incompleteness result and self-referential paradox is therefore very deep, since every proof of incompleteness has some version of self-referential paradoxicality lurking around in the background.

Through the system of encoding we have come to call "Gödel numbering" (of course self-effacing Gödel didn't name it so) G can be rendered in arithmetical notation, so that it also makes an arithmetical statement. Here is one of the places where the hard work comes in, and a bit later on in this chapter we'll dip a little deeper into this aspect of the proof. Gödel found an ingenious way of making an arithmetical language speak of its own formalism. The upshot of the technique is that G is simultaneously making two different statements, asserting an arithmetical claim and also asserting its own unprovability. In other words, what G has to say, in addition to its straightforward arithmetical content (it's going to be a weird arithmetical proposition, what with all the fiddlefaddling that gets us to it) is:

G *is unprovable in the system.*

The negation of G thus amounts to the proposition:

G *is provable in the system.*

If G were provable then its *negation*—which, after all, says that G is provable—would be true. But if the negation of a proposition is true, then the proposition itself is false. So if G is provable then it is false. But if G is provable, then it is also true. After all, what else does a proof show, assuming of course that the system is consistent (since in an *in*consistent system all propositions are provable). So, assuming the consistency of the system, if G is provable then it is both true and false—a contradiction—which means that G is *not* provable. Thus if the system is consistent, then G is not provable in it. But that is exactly what G says: that it isn't provable. So G is true. Therefore, G is both unprovable and true, which is precisely the famous conclusion of Gödel's proof, that there is a true but unprovable proposition expressible in the system if the system is consistent. And because G also has a straightforwardly arithmetical meaning which, of course, is true if G is true (because it *is* G) Gödel's proof shows that there are arithmetical truths (for example, G!) that cannot be proved within the formal system, assuming the system to be consistent. The formal system is either inconsistent or incomplete.

What is more, the proof demonstrates that should we try to remedy the incompleteness by explicitly adding G on as an axiom, thus creating a new, expanded formal system, then a counterpart to G can be constructed within that expanded system that is true but unprovable in the expanded system. The conclusion: There are provably unprovable, but nevertheless true, propositions in *any* formal system that contains elementary arithmetic, assuming that system to be consistent. A system rich enough to contain arithmetic cannot be both consistent and complete.

That's really the overall strategy. In some sense (once one assimilates the strange idea of a single proposition that can speak simultaneously of itself and of arithmetic), it is simple. Of course, the devil is all in the details; and it's to the diabolical details that we will now turn.

Step One: Lay Out a Formal System

Gödel begins his proof by laying out his formal system, which consists, as do all formal systems, of an alphabet of symbols, rules for combining these symbols into wffs, a special set of wffs called the "axioms," and a deductive apparatus for deriving (as "logical consequences") wffs from other wffs that must be either axioms or consequences of axioms.

In most systems of formal logic, it is convenient to have symbols for "and"(conjunction) and "or" (disjunction), as well as for the expressions "if . . . then . . ." (material implication) and ". . . if and only if . . .". There are also symbols that express the quantificational notions of "all" and "some." However, it will be more convenient to have as few basic symbols as possible. We can eliminate conjunctions in favor of disjunctions since "*p* and *q*" means the same thing as "it's not the case that *p* is false or *q* is false." So *I want to eat and I want to be thin* is equivalent to *It's not true that either I don't want to eat or don't want to be thin.* Then we can take our elimination one step further since disjunction can be eliminated in favor of material implication because "if *p*, then *q*" means the same thing as "not-*p* or *q*." We can also eliminate the notion of "some," by using the notion of "all" since "there are some *x*'s that are *F*" is the same as "it is not the case that all *x*'s are not *F*." So *Some logicians are rational* is equivalent to *Not all logicians are irrational.* After our eliminations, we're left with nine primitive concepts, and corresponding symbols, with which to express all of arithmetic in a formal system.

Step Two: Gödel Numbering

The next step in the proof is to devise a mechanical method of assigning a unique number to every proposition of the system. It's by assigning these numbers that the blending of the voices is accomplished, with arithmetical statements also making metamathematical statements.

The logician Simon Kochen beautifully described Gödel's proof to me as bearing a marked similarity to Kafka's work (which Gödel happened to admire). In both Kafka and Gödel, there is a certain Alice-in-Wonderland quality, a sense that one has entered a strange universe where things morph into others, including meanings themselves. Yet everything proceeds stepwise according to the most rigorous rule-bound logic. (The rigorous logic of Kafka is under-appreciated.) So much of the work in Gödel's proof amounts, in Kochen's words, to "bookkeeping." This double aspect of the proof mirrors something, Kochen also said, essential about Gödel's mind as well, the wild leaps of imagination coupled with a sort of legalistic ploddingness. Both of these aspects emerge in the Gödel numbering. Here's the basic idea:

The formal system, remember, involves a variety of types of objects: the symbols of the alphabet, the combinations of symbols into wffs, and special sequences of wffs (in other words, proofs). Everything is built out of the basic symbols of the alphabet: a wff is a sequence of these symbols, and a proof is a sequence of wffs (with the conclusion simply the last entry in the sequence).

The Gödel numbering thus begins by assigning each primitive symbol of the alphabet a number. Once each of the primitive symbols has a number, one continues with a rule for assigning numbers to the wffs themselves, based on the corresponding numbers of their constituents. Then, once we have a Gödel number for each wff, the Gödel numbering is completed with a rule for assigning numbers to sequences of wffs, which, of course, are what proofs are.

Furthermore, once every wff has been assigned a corresponding number, we will be able to analyze the structural relationships between the propositions by merely analyzing the arithmetical relationships between their corresponding numbers. Or vice versa. For example, one wff will be a consequence of another one precisely in case the Gödel numbers of the wffs are arithmetically related to each other in just the right way. In other words, two different sorts of descriptions will be collapsed into one another: arithmetical descriptions, setting forth relationships between numbers that are expressible within the formal system; and metadescriptions about the logical relationships holding between the wffs in the system. These metastatements are purely syntactic, as they are simply the consequences of the syntax of the formal system, that is, its rules.

The idea of Gödel numbering is basically the idea of encoding, which allows you to move back and forth between the original propositions and the code. In elementary school, my friends and I had a similar code we'd use to pass notes in class, assigning each letter of the alphabet a number between 1 and 26, with "A" assigned 1, "B" assigned 2, etc. "Meet me" was rendered as: 13 5 5 20 13 5.

The encoding system Gödel used had to ensure that the same Gödel number wouldn't be assigned to different things, say to both some wff as well as to a sequence of wffs (a proof). The rules for encoding give us an algorithm (which is, remember, a set of rules which tell us, at each step, how to proceed, often based on the result we get from a previous application of rules in the set) for getting from any wff, or sequence of wffs, to a unique Gödel number. There is also an algorithm for the reverse process: given any Gödel number, we can effectively figure out what formal object in the system it stands for. The Gödel numbering must obey one further condition: the translation of syntactic descriptions of the logical relationships between wffs in the system into arithmetical propositions that it provides must be such that these arithmetical propositions are themselves expressible within the system.

The spirit of Gödel numbering can be conveyed simply, though were we to be rigorous, as Gödel of course was, there would be nothing simple about the encoding. The modern positional notation is so familiar to us that we forget it is itself a coding system, requiring proof in the formal system. So, for example, the usual digital symbol 365 is shorthand for 3 times 10 squared, plus six times 10, plus five. Gödel's system of encoding used the exponential products of prime numbers and relied on the prime factorization theorem which states that every number can be uniquely factored into the product of primes. The prime factorization theorem ensured that there was an algorithm for getting from any wff or sequence of wffs to a Gödel number and vice versa. The digit juxtaposition we use below would, if made rigorous, be every bit as complicated as the system that Gödel used. But we won't be rigorous.

First, we'll arbitrarily assign a natural number to each symbol of the alphabet of the formal system. We can do all we need to do in setting out a formal system of arithmetic if we restrict ourselves to just nine symbols, to each of which we'll assign a Gödel number:

Basic sign	Gödel Number	Meaning
~	1	not
→	2	if . . . then . . .
x	3	variable
=	4	equals

Basic sign	Gödel Number	Meaning
0	5	zero
s	6	the successor of
(	7	punctuation mark
)	8	punctuation mark
′	9	prime

The quantificational notion of "all" is represented by using parentheses and variables. So, for example, $(x)F(x)$ means: all x's are F. The prime allows us to generate additional variables: x, x', x'', x''', etc. Since we have a symbol for 0, and a way of indicating successors, we have a way of indicating all the natural numbers.

We now specify a rule for assigning Gödel numbers to wffs, which of course are just sequences of the symbols of the alphabet. We adopt the easiest rule possible, reminiscent of the encoding my friends and I used in elementary school. We simply plug in the Gödel number for each symbol in the wff and that will be the Gödel number for the wff.

So consider this wff:

$$(\mathbf{p}_1) \qquad (x)\,(x')((s(x) = s(x')) \rightarrow (x = x'))$$

Assuming our universe of discourse (the objects to which the variables are interpreted to be referring) is the natural numbers, what p_1 says is that if two numbers have the same successor, then they're the same number. Or, in other words, that one number can't be the successor of two different numbers. More literally: *for all x and for all x′*, if the successor of x is identical to the successor of x', then x is identical to x'.

Now we're going to turn this wff into a number by just consecutively going along and replacing each symbol in the wff by its Gödel number. Every symbol in the formula p_1 has been assigned a number: the open parenthesis a 7, the *x* a 3, the close parenthesis an 8, the tilda a 1. Replacing each element of the formula with the corresponding Gödel number yields a large number, which is the Gödel number for the wff. Abbreviating "the Gödel number of the proposition p_1" by GN(p), we obtain:

$$GN(p_1) = 738739877673846739882734398$$

In this way Gödel numbers are assigned to wffs, which are sequences of symbols, and hence to propositions, which are just special wffs. In the same way, Gödel numbers can be assigned to sequences of propositions, and in particular to potential proofs, which are, after all, just sequences of propositions, using the Gödel numbers already assigned to propositions. Bookkeeping! In our simplified version, the Gödel number of a sequence of propositions (a potential proof) is obtained basically by putting the Gödel numbers of the sequenced propositions together; however, since it is important to be able to unambiguously extract from the number obtained the original sequence of propositions, we'll need some sort of signal that will indicate where one proposition ends and the next one begins—sort of like a carriage return on a typewriter. We'll let 0 function as our carriage return, indicating that we now go to a new line in the proof.

So, let us say that in a particular sequence of propositions, p_1 is followed by p_2, where p_2 is defined as:

(p_2) $$s(0) = s(0)$$

The Gödel number that will correspond to the sequence of p_1 followed by p_2 is

$$GN(p_1, p_2) = 738739877673846739882734398067584675 8$$

Through Gödel's inspired contrivances, all of the logical relations that hold between propositions in the formal system become arithmetic relations expressible in the arithmetical language of the system itself. This is the essence of the heart-stopping beauty of the whole thing. So if, for example, wff_1 logically entails wff_2, then $GN(wff_1)$ will bear some purely arithmetical relation to $GN(wff_2)$. Suppose, say, that it can be shown that if wff_1 logically entails wff_2, then $GN(wff_2)$ is a factor of $GN(wff_1)$. We would then have two ways of showing that wff_1 logically entails wff_2: we could use the rules of the formal system to deduce wff_2 from wff_1; or we could show that $GN(wff_1)$ can be obtained from $GN(wff_2)$ by multiplying by an integer. Suppose that $GN(wff_1) = 195589$ and $GN(wff_2) = 317$. 317 is a factor of 195589, since 317 multiplied by 617 = 195589. So that wff_1 logically entails wff_2 could be demonstrated either by using the formal rules of proof to arrive at wff_2 from wff_1, or, alternatively, by using the rules of arithmetic to arrive at $GN(wff_1) = 195589$ by multiplying 617 by $317 = GN(wff_2)$. The metasyntactic and the arithmetic collapse into one another.

Once one has this sort of collapse between logical implication and arithmetical relationship, we can go on and demonstrate that certain sequences of wffs—precisely those that constitute proofs—have an arithmetical property expressible in the system. Proofs, of course, are built up out of logical entailments. So deriving this arithmetical property (of the Gödel numbers of all and only the proofs in the system) is going to be a consequence of the sort of collapse discussed in the previous paragraph. The Gödel number of those strings of wffs that are valid proofs within the system will have some sort of arithmetical property, say, they'll all be even or they'll all be odd or they'll be prime numbers or the squares of primes or, more likely, a property a good deal more complicated. In other words, the metasyntactic relationship of provability will become an arithmetical relationship; that a string of wffs is a proof will be some property of numbers. From this it becomes possible to show that all and only the provable wffs in the system, the theorems, have a certain arithmetical property. You can see where we're heading: toward arithmetical propositions expressible in the system that also speak to the issue of their own provability within the system. The Gödel numbering allows some propositions to engage in an interesting sort of double-speak, saying something arithmetical and also commenting on their own situation within the formal system, saying whether they're provable.

The double-speak of these propositions could be compared to the sort of thing that sometimes happens within a dramatic play, in particular when the play presents actors as characters, with their own "real-life" relationships, and then presents these characters as actors in a play within the play. Through careful contrivance, the lines that the actors speak, in the play within the play, can also be interpreted as having a real-life meaning in their relationships outside the play within the play (in the play proper). Gödel's strategy asks us to grasp something analogous to what the country audience in Leoncavallo's opera *I Pagliacci* grasp when they understand the actors to be delivering lines that, as well as making sense within the play, have a meaning in their offstage lives (in the opera). In Gödel's inspired stage production, lines speak both to the formal relationships within the system, the play within the play, and also reveal real-life arithmetical relationships. The proposition that Gödel's proof constructs, the one that will simultaneously announce both its own (provable) unprovability and a true (unprovable) arithmetical relationship, has the same sort of double-entendre as Pagliacci's final tragic cry, "*La commedia è finita*!"—the comedy is over.

Step Three: Create a Proposition That's True Because It Says That It's Unprovable

Having established his ingenious layers of meaning, Gödel now conjures a rather amazing arithmetical property, which we'll designate as "Pr," for "provable." Because, although Pr is a purely arithmetical property, it's also that very property toward which all the contrivances have been heading. It's an arithmetical property that is true of all and only the Gödel numbers of the provable propositions of the system. I choose the verb "conjures" deliberately because even though the collapse of the metasyntactic and arithmetical was heading toward "Pr," there is still a sense of magic about the entrance of Pr.

Before getting to the property, one little technical point about the way in which properties are specified: by propositional functions of one variable. These can be thought of as of the form F(x). These are expressions in the formal system involving a single variable—the x which is a dummy standing for a whole range of possible values (the domain of individuals) that can be plugged in; if you plug in a value for the variable, you end up with a wff that is either true or false, in other words a proposition. F(x), just as it is, is neither true nor false. So say F(x) were to mean: *x is the successor of 1*. This is neither true nor false; it isn't a proposition, a definite statement, since what it says depends upon what x actually represents. Plugging in 2 for the variable x yields a true proposition, and plugging in anything else yields false propositions. That's how properties are designated in the system, by propositional functions of one variable.

Now onward to $\mathrm{Pr}(x)$, which is a somewhat complicated property of numbers. Remember, first of all, that each of the wffs in the formal system have been assigned numbers through the wonders of Gödel numbering. So for every wff p, we have some $\mathrm{GN}(p)$, some natural number or other. The theorems are a special subset of the wffs of the system, the provable propositions. Given any natural number n, then, it may or may not be the case that it corresponds to some theorem in the formal system, that is, it may or may not be the case that $n = \mathrm{GN}(p)$ for some proposition p which is a theorem of the formal system.

Now we are in position to define $\mathrm{Pr}(x)$. The possible values for this propositional function, the things we plug in for the dummy x, are (the expressions for) the natural numbers. For any natural number n, if $n = \mathrm{GN}(p)$ for a theorem p of the system then we say that n satisfies the property $\mathrm{Pr}(x)$, that is, that $\mathrm{Pr}(n)$ is true. Gödel showed that this property is one that can in fact be expressed in the formal system, that is, that it's an example of an $F(x)$. $\mathrm{Pr}(x)$ is a formally expressible arithmetical property, albeit one that is extremely complicated, not anything that we can explicitly give here. But by means of this property Gödel is able to take metasentences about the system, stating which propositions are theorems of the system, and transform them into arithmetical sentences within the system: "p is a theorem" is transformed into "$\mathrm{Pr}(\mathrm{GN}(p))$." To say of some particular n

that it has the property Pr(x) is to say that this number corresponds to a theorem in the formal system.

Now perhaps you're thinking something like this: That a particular number n has the property Pr(x) isn't a *real* property of n. The number n, for example, may be even or it may be odd; if it's divisible by 2 then it's even. Let's say it is. Then its evenness is a real arithmetical property; n couldn't be the number it is if it weren't even. In contrast, looking at this property Pr(x) metasystematically, it doesn't seem at all a proper sort of numerical property. A number n has this property only arbitrarily, since the proposition with which n is associated is arbitrary, a consequence merely of the way in which the Gödel numbering was set up, the inspired contrivances.[6] True enough, but, arbitrary though it may be, Pr(x) is nonetheless a real arithmetical property, and a number n either will or won't have it. And n wouldn't be the number that it is unless it either did or didn't have it. Just because Pr(x) has a metameaning doesn't detract from its arithmetical character. This property Pr(x) is, in fact, stupendous, and it allows us to take the plunge into the heart of the proof.

What we use next is something called the diagonal lemma. It's a general statement, a particular case of which is what we need to prove Gödel's theorem. (Gödel didn't actually use it, but rather derived the particular case.) Making use of this

6 Compare this to Richard's päradox (footnote 5 above), which Gödel cites, together with the liar's paradox, as a heuristic aid to understanding his proof. Richard's paradox also has the (illusive) feel of attributing a contrived, or unreal, property to numbers (the property of being Richardian) which a number will or won't have because of the arbitrary assignments of numbers to properties.

general lemma (which, of course, we're not going to prove) will simplify matters enormously.

The diagonal lemma states that the Gödel numbering is such that for any propositional function F(x) of one variable, there exists a number n such that the Gödel number of F(n), the proposition we get when we plug n into the function F(x), turns out to be n itself. (That there must exist such a number for each F(x) may suggest what superhuman efforts went into the Gödel numbering.) In other words, the diagonal lemma asserts that for any F(x) there is an n such that

$$n = \mathrm{GN}(\mathrm{F}(n)) \qquad \textbf{(0)}$$

The number you get out is the same number you started with, and so the special n associated with a given F corresponds precisely to where the graph of $y = \mathrm{GN}(\mathrm{F}(x))$ intersects the diagonal, the graph of $y = x$, namely where $x = n$. Hence the name, the "diagonal lemma."

(N.B. The statement $n = \mathrm{GN}(\mathrm{F}(n))$ is in the metalanguage. It's a "normal statement" (i.e., not a formal statement) and n on the left denotes a (normal) natural number. However, the n on the right represents the expression in the formal system that represents the number n, namely s(s(s(. . . s(s(0))))), with n occurrences of s.)

Notice that the number n that the diagonal lemma associates with the propositional function F(x) is such that the proposition with Gödel number n (namely F(n)) says that n itself has the property F. It is, roughly speaking, of the form: this very sentence is F. Soft whispers of self-referentiality are hovering in the hushed air.

Now let's go back to that stupendous property that Gödel produced, Pr, which is true of all and only the Gödel numbers of the theorems, the provable wffs, of the system. A number will have the arithmetical property Pr if and only if it corresponds to a provable proposition under the Gödel numbering. In other words:

$$\mathrm{Pr}(\mathrm{GN}(p)) \text{ if and only if } p \text{ is provable}$$

We have the metasyntactic sense of what $\mathrm{Pr}(x)$ means (though not what its arithmetical meaning is; you just have my assurance that it has an arithmetical meaning). Now let's look at the property:

$$\sim \mathrm{Pr}(x)$$

This second property will be true of all and only the Gödel numbers of nontheorems; that is, it is true of the Gödel numbers of objects that are not propositions provable in the system. In other words:

$$\sim \mathrm{Pr}\,(\mathrm{GN}(p)) \text{ if and only if } p \text{ is not provable} \qquad (1)$$

What kind of sentence is (1)? It is a metamathematical sentence. It's not itself a sentence of the formal system, nor an arithmetical sentence. But by means of (1) we can transform some metamathematical sentences into arithmetical sentences.

Now the diagonal lemma concerns *any* propositional function $F(x)$, so let's apply it to $F(x) = \sim\mathrm{Pr}(x)$. The diagonal lemma, remember, states that for any propositional function $F(x)$ of one variable, there exists a number n that is the Gödel number of the very proposition we get when we plug n itself into the function. We're now considering the propositional function $\sim$Pr. According to the diagonal lemma there is some number, let's call it g, such that:

$$g = GN(\sim Pr(g)) \qquad (2)$$

We arrived at (2) by replacing F by ~Pr and *n* by g in (0). Equation (2) says that g is the Gödel number of the proposition that states that the number g lacks the arithmetical property Pr (which belongs to all and only those numbers which are Gödel numbers of provable propositions).

Now we're ready to frame our proposition G. Let

$$G = \sim Pr(g) \qquad (3)$$

The proposition G states that the number g lacks the property Pr. Moreover, by (2) and (3), we have that:

$$GN(G) = g$$

Now we go back to (1) and make some substitutions. Here's (1) again:

$$\sim Pr\,(GN(p)) \text{ if and only if } p \text{ is not provable}$$

Letting p = G in (1), and using the fact that GN(G) = g, we obtain:

$$\sim Pr(g) \text{ if and only if G is not provable} \qquad (4)$$

That is, using again (3), G if and only if G is not provable. What (4) is saying is that G is true if and only if G is not provable!

G is, of course, a purely arithmetical statement, but it's also simultaneously talking about itself, and what it's saying is that it's not provable. Is what it's saying true? Well, it could hardly be false since then it would have to be provable and hence true anyway. That is, unless, of course, the formal system of arithmetic is inconsistent, so that all its propositions would be provable, even contradictions. This is the point in the proof where that presumption of the consistency of the formal system is being called upon to stand up and deliver. And it does. G is both unprovable and, given that *that's what it says*, it's also true. We haven't shown that it's true by finding a proof for it within the formal system, using the purely mechanical rules of that system, that is, by deducing it. Rather, we've shown it's true by, ironically, going outside the system and showing that no proof for it *can* be produced within the formal system. We've shown that G is true by showing that it can't be proved, just as it says.

Moreover, Gödel in fact showed how to construct a true but unprovable proposition, not just for the formal system of arithmetic we have been discussing but for any formal system whatsoever containing arithmetic. So if we should try to weasel our way out of Gödel's first incompleteness theorem by constructing a new formal system that has G appended as an axiom, a new problematic proposition can be constructed for that system. And so on, ad infinitum. There are provably unprovable but nonetheless true propositions in any formal system that contains elementary arithmetic, assuming that system to be consistent.

And that is Gödel's first incompleteness theorem.

The Second Incompleteness Theorem

The second incompleteness theorem states that the consistency of a formal system that contains arithmetic can't be formally proved within that system. It would seem to follow, pretty straightforwardly, from the first. Remember that the first incompleteness result has the form of a conditional statement: if the formal system of arithmetic is consistent, then G is unprovable. Let C stand for the proposition: "The formal system of arithmetic is consistent." So the first incompleteness theorem tells us: if C, then G is unprovable. The arithmetization of the proposition "G is unprovable" is, of course, G. So what the first incompleteness says is: C → G,.and this conclusion was proved within the formal system of arithmetic. So if we can go on and prove C within the formal system of arithmetic, we would ipso facto be proving G within the formal system of arithmetic, since we've proved C → G. And since G has been proved unprovable within the formal system of arithmetic, we know that C, too, is unprovable in the formal system of arithmetic.

And that is Gödel's second incompleteness theorem.

Notice, by the way, that the second incompleteness theorem doesn't say that the consistency of a formal system of arithmetic is unprovable by any means whatsoever. It simply says that a formal system that contains arithmetic can't prove the consistency of itself.[7] The formal system of arithmetic is clearly consistent, one might want to argue, if we avail ourselves of semantic considerations. After all, the natural numbers constitute a model of the formal system of arithmetic, and if a system has a model then it's consistent. Recall the lessons learned from the description of my New York apartment. So long as I describe it unambiguously and accurately (it has, for example, one bathroom), I don't need to worry

7 In 1936 Gerhard Gentzen, a member of the Hilbert school, proved the consistency of arithmetic, but it wasn't within a finitary formal system. His proof involved the sort of transfinite reasoning that Hilbert had proposed be banished in favor of finitary formal systems.

about self-contradictions (it has four bathrooms) lurking in the description.) In other words, when the formal system of arithmetic is endowed with the usual meanings, involving the natural numbers and their properties, the axioms and all that follow from them are true and, therefore, consistent. This sort of argument for consistency, however, goes outside the formal system, making an appeal to the existence of the natural numbers as a model. This is not the kind of reasoning to offer solace to a formalist, however much it might gladden the heart of a Platonist like Gödel. Finitary formal systems were, according to Hilbert, *anschaulich*, transparently pure. With everything reduced to either stipulation or the logical consequences of the stipulated mechanical procedures, there is no place for obscurity (possibly infected with paradox) to creep in. Finitary formal systems were, according to Hilbert's program, the means for draining the paradoxical substance out of the notion of the infinite:

> Operating with the infinite can be made certain only by the finite. The role that remains for the infinite to play is purely that of an idea—if one means by an idea, in Kant's terminology, a concept of reason which transcends all experience and which completes the concrete as a totality—that of an idea, moreover, which we may unhesitatingly trust within the framework erected by our theory.

Hilbert's program—to expunge all reference to intuitions—was most particularly directed toward our intuitions of infinity; not surprisingly, finite creatures that we are, it is these intuitions that have proved themselves, from the very beginning, to be the most problematic. The deep uneasiness that even Euclid felt toward his fifth postulate and which was replicated down through the ages, stems from our altogether appropriate lack of confidence on all matters infinite. Yet one can't do any mathematics at all, not even basic arithmetic, without referring implicitly to the infinite. If we could tame infinitude by capturing it within our finitary formal systems then we would have effected the perfect compromise. The infinite would, in Hilbert's words, "be made certain by the finite."

Gödel's result, in effect, proclaims the robustness of the mathematical notion of infinity; it can't be drained of its vitality and turned into a ghostly Kantian-type idea hovering somewhere over, but without entering into, mathematics. The mathematician's intuitions of infinity—in particular, the infinite structure that is the natural numbers—can no more be reduced to finitary formal systems than they can be expunged from mathematics.

Another way of seeing the robustness of our intuition of infinity is to consider that it follows from Gödel's work that there are "nonstandard models" of arithmetic. Specifically, Gödel's completeness theorem—that Ph.D. thesis that turned out to be not as humdrum as it had first seemed—tells us, among other things, that every consistent formal system has a model. There is a way of specifying a universe of discourse and interpreting the predicates and relations and constants so that all the theorems of the formal system are true descriptions. It follows from this, together with Gödel's first incompleteness theorem, that there is at least one nonstandard model of arithmetic: a model that satisfies all the axioms of the formal system of arithmetic but in which some of the truths of standard arithmetic—G, for example—will be false.

So this nonstandard model isn't going to consist of the natural numbers as we know and love them.[8]

The natural numbers transcend the formal system of arithmetic, in the sense that that formal system does not uniquely pick out the natural numbers as its model; as being, that is, what the formal system is about. The same thing happens to be true of any larger formal system containing arithmetic. There remains something—always—that eludes capture in a formal system. It was in this metalight that Gödel viewed his incompleteness theorems.

Gödel's second incompleteness theorem, the direct consequence of the first, as von Neumann was quick to realize, effectively demolished Hilbert's program for mathematical transparency, since finitary formal systems could only be proved consistent by resorting to arguments that couldn't be expressed within the formal systems themselves, no matter how they were modified and extended.

The second incompleteness theorem put formalism in an impossible bind: the formalist incentive was to banish the opacity of the nature of the thing in itself (space, numbers, sets) for

8 The study of model theory—interpretations both standard (with the natural numbers as the universe of discourse) and nonstandard—as distinct from proof theory—the study of the purely syntactic features of formal systems—was opened up as a result of Gödel's incompleteness proof. Not only had Gödel's proof put logic, in the words of Simon Kochen, "on the mathematical map," but it also had pointed the way to new and distinct regions of technical research. Gödel himself never showed much interest in doing research in the areas his proof engendered, not so surprising in light of his audacious ambition to restrict himself only to mathematics with metamathematical implications.

the transparency of formal systems. But it's of the highest priority that a formal system—drained of the descriptive content that would, so long as the axioms were truly descriptive, ensure its consistency—be proved consistent. This can only be done by going outside the formal system and making an appeal to intuitions that can't themselves be formalized. (The last article that Gödel was to publish in his life showed how arithmetic could be shown to be consistent provided one makes certain assumptions about objective mathematical reality.) The purely syntactic aspects of formal systems—the transparent aspects—aren't sufficient unto themselves, neither in being able to prove all the true arithmetical propositions expressible within the system (the first incompleteness theorem) nor in providing a proof of internal consistency (the second incompleteness theorem).

What was Hilbert's reaction to the logical wrench thrown into his beautiful plan? The mathematician, Paul Bernays (1888–1977), who had come to Göttingen to serve as Hilbert's assistant, reveals in a letter that he had, sometime before Gödel's proof, himself become "doubtful . . . about the completeness of the formal systems" and had "uttered [his doubts] to Hilbert." Hilbert, according to Bernays, became angry; and he was angry when Gödel's proof became known to him.

But a proof is a proof, as Hilbert, of all people, appreciated.

Wittgenstein and Incompleteness

Wittgenstein's reaction to Gödel's proof was notably different from Hilbert's. He did not live to accommodate himself to Gödel's work, as Hilbert did, no matter how unpalatable to his philosophical outlook, to his entire program, it was. There is a logical incompatibility between Wittgenstein's views on the foundations of mathematics (both early, positivist-sounding and later postmodernist-sounding Wittgenstein) and Gödel's incompleteness theorems. Wittgenstein acknowledged the incompatibility; he did not, like so many others, take the mathematical results and bend them into some metamathematical shape more to his liking, the shape of positivism or existentialism or postmodernism. He acknowledged the incompatibility and countered that Gödel therefore could not have proved what he thought he had proved:

> Mathematics cannot be incomplete; any more than a *sense* can be incomplete. Whatever I can understand, I must completely understand. This ties up with the fact that my language is in order just as it stands, and that logical analysis does not have to add anything to the sense present in my propositions in order to arrive at complete clarity.

It is really not so surprising that Wittgenstein would dismiss Gödel's result with a belittling description like "*logische Kunststücke*," logical conjuring tricks, patently devoid of the large metamathematical import that Gödel and other mathematicians presumed his theorems had. Gödel's proof, the very possibility of a proof of its kind, is forbidden on the grounds of Wittgensteinian tenets that remained constant through the transformation from "early" to "later" Wittgenstein, where early Wittgenstein had a monolithic view of language and its rules and later Wittgenstein fractured language into self-contained language-games, each functioning according to its own set of rules. He was adamant on the impossibility of being able to speak about a formal language in the way that Gödel's proof does. He was also adamant in denying that paradoxes, being trivial epiphenomena of the ways in which language works, could have large and interesting consequences. (He was to argue this very point with the logician Alan Turing, who ignored Wittgenstein and went on to produce another extraordinary proof which shares many attributes with Gödel's; so many, in fact, that it yields an alternative proof for the incompleteness of formal systems rich enough to express arithmetic.) He was, more generally, adamant in denying that mathematical results, being the results of mere syntax, could have large and extra-mathematically interesting consequences: "No calculus can decide a philosophical problem. A calculus cannot give us information about the foundations of mathematics." He was, in short, adamant in denying the possibility of a proof such as Gödel's.

All of which adamant irreconcilabilities provide context for understanding a statement of Wittgenstein's that has tended to irritate mathematicians: "My task is not to talk about Gödel's proof, for example. But to by-pass it." Yet Wittgenstein does circle back, again and again, in his *Remarks on the Foundations of Mathematics*, to Gödel's incompleteness theorem, deconstructing it, as the postmodernists would say, trying to show that its meaning is at odds with its intent, that it cannot mean what it purports to mean.

Yet, if one pushes beyond the metamathematical irreconcilabilities separating Wittgenstein and Gödel, one comes upon a surprising commonality, at least between the early Wittgenstein and the logician, masked by the positivist interpretation of Wittgenstein. In a sense, the early Wittgenstein put forth an incompleteness thesis of his own in his final proposition of the *Tractatus*. Just as Gödel demonstrated that our formal systems cannot exhaust all that there is to mathematical reality, so the early Wittgenstein argued that our linguistic systems cannot exhaust all that there is to non-mathematical reality. All that can be said can be said clearly, according to the *Tractatus;* but we cannot say the most important things. We cannot speak the unspeakable truths, but they exist. Again, we see why Wittgenstein fulminated against the positivists, why he sometimes became so enraged with his would-be disciples that he turned his back to them and faced the wall, reciting the poetry of a mystic Indian (an act hostile to positivists, if ever there was one: curious that its latent hostility seems to have passed them by).

For Gödel, for each formal system there will be truths expressible in that system that will not be provable; and one of the most important truths about the system, that it is consistent, will not be provable within the system. So both Gödel and the early Wittgenstein are united against the positivists' reiteration of the ancient Sophist's slogan that man is the measure of all things. Both men assert a fundamental incompleteness that takes the measure of man.

Wittgenstein's is, by far, the more radical statement of incompleteness. For Gödel's there is expressible knowledge which cannot be formalized. The limits of formalization, of our attempt to reduce all mathematical knowledge to the specified rules of a system, are not congruent with the limits of our knowledge. Our mathematical knowledge exceeds our systems. For early Wittgenstein there is no expressible knowledge that escapes the limits he delineates. On the *other* side of meaningfulness lies all the most important subjects: ethics and aesthetics and the meaning of life itself. "There are, indeed, things that cannot be put into words. They make themselves manifest. They are what is mystical."

Wittgenstein's unpositivistically positive attitude toward the idea of the mystical—even though it is the meaningless mystical—might have struck a responsive chord in Gödel. Gödel was even receptive to the suggestion that *his* incompleteness theorems had consequences in the mystical, or at least religious, sphere. In a letter to his mother on 20 October 1963 he remarked with regard to an article that she had sent him, and which he had not yet read, concerning the implications of his work: "It was something to be expected that sooner or later my proof will be made useful for religion, since that is doubtless also justified in a certain sense." At the very least, Gödel believed his first incompleteness theorem supported Platonism's insistence on the existence of a suprasensible domain of eternal verities. Platonism isn't of course tantamount to religion or mysticism, but there are affinities.

For early Wittgenstein, as for Gödel, the attempt to systematize reality, to capture it all within our limpid constructions designed to keep out all contradictions and paradox, are doomed to failure. Gödel's first incompleteness theorem tells us that any consistent formal system adequate for the expression of arithmetic must leave out much of mathematical reality, and his second theorem tells us that no such formal system can even prove itself to be self-consistent. Of course, Gödel believes that these systems *are* consistent, since they have a model in the truly existent abstract realm. Wittgenstein so ardently embraces the futility of attaining both completeness and self-consistency that he allows the *Tractatus* itself to bare its self-contradiction in plain sight, speaking of that of which one cannot speak, even while pronouncing the very statement that forbids it.

Gödel would most likely not have known that, on some level, he and (the early) Wittgenstein shared a profound conviction of incompleteness, a shared rejection of the logical positivists' endorsement of the Sophist's "measure of all things." After all, as he reported on his Grandjean questionnaire, he never studied Wittgenstein for himself. His acquaintance with the philosopher was, by his own estimation, superficial, presumably because he was not sufficiently excited by what he heard to study the philosopher for himself; he knew only what he learned by way of the discussions of the Vienna Circle. And the logical positivists, studying the precisely obscure *Tractatus* proposition by proposition, were intent on systematically ignoring those aspects that would have been congenial to Gödel, speaking to his own conviction of a reality always escaping our ordered attempts at precision.

Of course, Gödel and Wittgenstein located the escaped parts of reality in irreconcilably different ways. Gödel's conviction, the metamathematical interpretation he gave his incompleteness theorems (as well as his work on the continuum hypothesis), was that it was aspects of mathematical reality that must escape our formal systematizing (although not our knowledge), and Wittgenstein's view on the foundations of mathematics would not countenance this conviction. For Wittgenstein, at least early Wittgenstein, all of knowledge, a fortiori mathematical knowledge, is systematizable; what systematically escapes our systems is the unsayable, which includes all that is important. Gödel believed our expressible knowledge, demonstrably our mathematical knowledge, is greater than our systems. Whereof we cannot formalize, thereof we can still know, the mathematician might have said, had he had any inclination toward the oracular.

Wittgenstein never allowed Gödel's result to tamper with his views on metamathematics, which subject increasingly obsessed him in the years after the *Tractatus* was published and subsequently renounced by its author. His aim, as he said, was to bypass Gödel's proof. This is both interesting in itself and interesting because of its galling effect on Gödel. The philosopher had spoken of necessary silence. Gödel, one suspects, would have liked that silence to envelope the philosopher himself.

The Spreading Incompleteness

The incompleteness proof opened up entirely new areas of research, most notably model theory and recursion theory. Gödel was never interested in pursuing the problems of those fields for himself. Much like his soulmate Einstein, he was interested in pursuing what Einstein called problems of "genuine importance," that is problems that lay in the interstice between exact science and philosophy, problems that radiated meta-implications. He left the "mop-up work" (in Thomas Kuhn's colorful terminology) to others. The vaulting intellectual ambition and confidence—so incongruously coupled with worldly fearfulness and self-effacement—may have meant that he left behind fewer results than he might have, but it also meant that the reach of his results was vast.

Gödel's incompleteness theorems do not stun us simply because they open up promising new areas of technical research. Deep discoveries in the exact sciences quite often do exactly that. What makes Gödel's results so stunning is the sheer volume of all that they have to say. The passionate Platonist, who had sat mum among the positivists, not murmuring a word of demurral, had produced the most loquacious theorems in the history of mathematics.

It was because of their *volubility* that a philosopher like

Wittgenstein could not accept them, that his self-assigned task was not to discuss them but to bypass them.

(Wittgenstein was to carry on his extended argument against the very possibility of such a result as Gödel's with the young English logician Alan Turing, who would go on to produce a proof that has great affinity with Gödel's. Turing, too, would manage to give sharp mathematical expression to metamathematical concepts and to appropriate the structure of self-referential paradoxicality to his own ends. Alan Turing had spent the academic year 1936–37 at the Institute for Advanced Study, where Gödel's incompleteness theorems were very much the topic of the day among von Neumann and his circle. (Von Neumann did more than anyone else to disseminate the news of Gödel's accomplishments.[9]) Turing returned to Cambridge, with his mind dwelling on Gödel's proof. His first semester back in England, he gave a course in Cambridge on the "Foundations of Mathematics." That same semester Wittgenstein was also giving a course there entitled "Founda-

9 For example, the logician Stephen Kleene recounted how Gödel "entered my intellectual life. . . . One day in the fall of 1931 the speaker in the mathematics colloquium at Princeton was John von Neumann. Instead of talking on work of his own (of which there was plenty) he spoke on the results of Gödel's *1931* paper, which had recently come out in the *Monatshefte*, but which Church and we in his course had not yet noticed. Von Neumann had had a preview of the first of those truths (accompanied by intellectual intercourse with Gödel) at the Königsberg meeting of September 1930. After the colloquium, Church's course continued uninterruptedly concentrating on his formal system; but on the side we read Gödel's paper, which to me opened up a whole new world of fascinating ideas and perspective. The impression this made on me was so much the greater because of the conciseness and incisiveness of Gödel's treatment."

tions of Mathematics," but the two courses could not have been more dissimilar. While Turing's course was, in effect, an introduction to mathematical logic, Wittgenstein devoted his course primarily to arguing against the possibility of mathematical logic in general, and against its implications for metamathematics in particular. Turing attended Wittgenstein's lectures, at least for a while, and Wittgenstein was so intent on changing Turing's mind that, while the logician attended, Wittgenstein's lecture was focused entirely on that aim; when Turing once mentioned that he would not be able to attend the seminar the next week, Wittgenstein remarked that then the discussion that week would be "parenthetical."[10] Eventually

10 The crux of the raging debate carried on between Turing and Wittgenstein that semester was whether contradictions and paradoxes can have any significance. Wittgenstein maintained that they cannot. Take, for example, the case of the liar's paradox. Wittgenstein's view about it was: "It is very queer in a way that this should have puzzled anyone—much more extraordinary than you might think: that this would be the thing to worry human beings. Because the thing works like this: if a man says 'I am lying' we say that it follows that he is not lying, from which it follows that he is lying and so on. Well, so what? You can go on like that until you are black in the face. Why not? It doesn't matter." But Turing, being committed to mathematical logic and aware of the use to which Gödel had put traditional paradoxes such as the liar's paradox, was very much under the impression that the liar's paradox—that paradoxes and contradictions in general—*do* matter, and that they are sometimes pointing the way to almost *necessarily* surprising truths. When Wittgenstein remained adamant that a contradiction in a system was no cause for concern, since everything reduced ultimately to the arbirtrariness of language-games, Turing stopped attending the lectures. Soon after this, Turing produced his metamathematical proof. Where Gödel had subjected the concepts of "provability" and "completeness" to his transformative techniques, Turing would give mathematical expression to the concepts of "decidability" and "computability." A mathematical question of a certain type

Turing stopped attending, and soon after produced his own important metamathematical result.)

(that includes an infinite number of specific questions) is *decidable* if and only if there exists an algorithm—one single computable series of operations—for determining, for any such question, whether the answer is yes or no, without necessarily explaining why the answer is yes or no. You don't have to understand why an algorithm works for it to be an algorithm and for it to work. In particular, there is the sort of mathematical question that asks whether or not a proposition is formally provable. It's not hard to see, though a little beyond the scope of this footnote, that Hilbert's formalism—or more precisely the notion, shown to be false by Gödel, that for every mathematical proposition either it or its negation admits of a proof—implies that the question of whether a proposition is provable is in fact decidable. If one had an algorithm for showing whether a proposition or its negation was provable, then, given Hilbert's formalism, one would have an algorithm for mathematical truth. Such an algorithm would provide a purely finitary combinatoric method for capturing the concept of mathematical truth (just as the concept of mathematical provability had been captured). Turing proved that no such algorithm exists, providing yet another frustration to Hibert's formalist hope. His proof is so closely allied with Gödel's that it is possible to derive an alternative proof of Gödel's first incompleteness theorem from it. Gödel was so gladdened by Turing's work that, in 1963, when his famous paper of 1931 was republished, he appended a paragraph stating that his own two incompleteness theorems had been strengthened by Turing's work. "In consequence of later advances, in particular of the fact that due to A.M. Turing's work (Turing [1937] 'On computable numbers, with an application to the Entscheidungsproblem,' *Proceedings of the London Mathematical Society*, 2nd series, 42, 230-65) a precise and unquestionably adequate definition of the general notion of formal system can now be given, a complete general version of Theorems VI and XI is now possible. That is, it can be proved rigorously that in *every* consistent formal system that contains a certain amount of finitary number theory there exist undecidable arithmetic propositions and that, moreover, the consistency of any such system cannot be proved in the system." Unfortunately, Turing and Gödel never met. Turing died at the age of 42, a suicide.

And it was because of their *volubility* that Hilbert's program was abandoned. Hilbert had tried to inoculate mathematics against paradox by eliminating all appeals to intuition. Gödel had proved that appeals to intuition could not be eliminated; he had undermined formalism's inoculation program. In particular, our intuitions about infinity—as susceptible as they are to invalidating fallacies, as the infamous paradoxes make all too clear (and which we can only avoid by adopting such ad hoc rules as Russell and Whitehead had devised)—nonetheless cannot be replaced by the semantics-free mechanical processes of mindless symbol-manipulation.

Such metamathematical conclusions, emerging from an a priori mathematical proof, are extraordinary enough in themselves. If these metamathematical results constituted all that followed from Gödel's incompleteness theorems it would still be sufficient to mark his work as singularly gabby. But Gödel's incompleteness theorems have been heard as addressing, in their irrepressible effusiveness, issues that range far beyond even metamathematics. Eminent thinkers have interpreted the incompleteness theorems as having something to say on the central question of the humanities, viz. *what is it that makes us human?* For mathematical theorems to have anything at all to say on such a subject as this—embedded deep in the messy matter of the human predicament—is to take what is already extraordinary and raise it to an altogether higher order of astoundingness.

The formalists had tried to certify mathematical certitude by eliminating intuitions. Gödel had shown that mathematics cannot proceed without them. Restricting ourselves to formal syntactic considerations will not even secure consistency. But these mathematical intuitions that cannot be eliminated and cannot be formalized: what *are* they? How do they come to be available to the likes of us? We are once again thrown up against the mysterious nature of mathematical knowledge, against the mysterious nature of *ourselves* as knowers of mathematics. How do we come to have the knowledge that we do? How *can* we? Plato himself had argued that the very fact that our reasoning mind can come into contact with the eternal realm of abstraction suggests that there is something of the eternal in us: that the part of ourselves that can know mathematics is the part that will survive our bodily death. Spinoza was to argue along similar lines.

Few scientifically minded, post-Gödel thinkers would perhaps be ready to follow Plato and Spinoza into drawing conclusions of our immortality from our capacity for mathematical knowledge. After all, we are not only living with the truth of Gödel but also the truth of Darwin. Our minds are the products of the blind mechanism of evolution. Still, many scientifically minded, post-Gödel thinkers have testified to hearing, within the strange music of Gödel's mathematical theorems, tidings about our essential human nature. They have argued from Gödel's incompleteness theorems to conclusions about what we are; or rather, to be more precise, about what we are *not*. Gödel's theorems tell us, according to this line of reasoning, *what our minds simply could not be.*

In particular, what our minds could not be, so goes the reasoning, are computers. The mathematical knowledge that we possess cannot be captured in a formal system. That is what Gödel's first incompleteness theorem seems to tell us. But formal systems are precisely what captures the computing of computers, which is why they are able to figure things out without having any recourse to meanings. Computers run according to algorithms and we, it seems, do not, from which it straightforwardly follows that our minds are not computers.

The first of the arguments claiming a connection between Gödel's first incompleteness theorem and the nature of the mind was published in 1961 by the Oxford philosopher John Lucas:

> Gödel's theorem seems to me to prove that Mechanism is false, that is, that minds cannot be explained as machines. So also has it seemed to many other people: almost every mathematical logician I have put the matter to has confessed to similar thoughts, but has felt reluctant to commit himself definitely until he could see the whole argument set out, with all objections fully stated and properly met. This I attempt to do.

Lucas's argument was stalwartly straightforward. No matter how complicated a "thinking" machine we engineer, he argued, this machine will run according to hard-wired rules that can be stated in a formal system, and when we ask this machine to tell us what the true propositions are it will be able to do this only by seeing which propositions follow according to the rules of the system. There will therefore be a proposition that eludes its grasp of truth, which is nothing but rule-determined provability—a proposition that our minds will nonetheless be able to grasp as true. No matter how we strengthen the machine, by adding in the previously elusive propositions as axioms, there will be yet another proposition that will elude it . . . but not us:

> This formula the machine will be unable to produce as being true, although a mind can see that it is true. And so

> the machine will still not be an adequate model of the mind. We are trying to produce a model of the mind which is mechanical—which is essentially "dead" —but the mind, being in fact "alive," can always go one better than any formal, ossified, dead system can. Thanks to Gödel's theorem, the mind always has the last word.

The mathematician Roger Penrose, also an Oxford don, has published two books, *The Emperor's New Mind* and *Shadows of the Mind,* arguing the case that Gödel's incompleteness theorems entail the falsity of mechanism, the deadendedness of the field of artificial intelligence, if artificial intelligence presumes to fully explain our thinking. His argument is much the same as Lucas's, though he does an even more thorough job of trying to anticipate and answer all possible objections.

> What did Gödel's theorem achieve? It was in 1930 that the brilliant young mathematician Kurt Gödel startled a group of the world's leading mathematicians and logicians, at a meeting in Königsberg, with what was to become the famous theorem. It rapidly became accepted as being a fundamental contribution to the foundations of mathematics—probably the most fundamental ever to be found—but I shall be arguing that in establishing his theorem, he also initiated a major step forward in philosophy of mind.
>
> Among the things that Gödel indisputably established was that no *formal system* of sound mathematical rules of proof can ever suffice, even in principle, to establish all the true propositions of ordinary arithmetic. This is certainly remarkable enough. But a powerful case can also be made that his results showed something more than this, and

> established that human understanding and insight cannot be reduced to any set of rules. It will be part of my purpose here to try to convince the reader that Gödel's theorem indeed shows this, and provides the foundation of my argument that there must be more to human thinking than can ever be achieved by a computer, in the sense that we understand the term "computer" today.

Penrose believes that even though the mind is not a computer, it is nevertheless a physical system. The mind is identical with the brain. Therefore, the nonmechanistic nature of the mind following, he claims, from Gödel's first incompleteness theorem, should direct our thinking toward nonmechanistic physical laws of just such a sort as are suggested by quantum mechanics. The mathematically intuiting mind, which demonstrably can't be captured mechanistically, is nonetheless a physical system; we should, therefore, look toward developing a nonmechanistic, radically new sort of science—the mysteries of quantum mechanics should be our guide here—so that the noncomputational aspects of mind can be accommodated. The noncombinatorial but nevertheless *physical* nature of thinking shows us the noncombinatorial nature of basic physical laws.

Gödel himself was far more reserved about drawing conclusions concerning the nature of the human mind from his famous mathematical theorems. What is rigorously proved, he suggested in his conversations with Hao Wang as well as in the Gibbs lecture that he gave in Providence, Rhode Island, 26 February 1951 (which he never published), is not a categorical proposition as regards the mind. Rather what follows is a disjunction, an "either-or" sort of a proposition. That is, he was admitting that nonmechanism doesn't follow, clean and simply, from his incompleteness theorem. There are possible outs for the mechanist.

According to Wang, Gödel believed that what had been rigorously proved, presumably on the basis of the incompleteness theorem, is: "Either the human mind surpasses all machines (to be more precise it can decide more number theoretical questions than any machine) or else there exist number theoretical questions undecidable for the human mind."

What exactly did Gödel have in mind with this second disjunct? I *think* that what he is considering here is the possibility that we are indeed machines—that is, that all of our thinking is mechanical, determined by hard-wired rules—but that we are under the *delusion* that we have access to unformalizable mathematical truth. We could possibly be machines who suffer from delusions of mathematical grandeur. What follows from his theorem, he seems to be suggesting, is that just so long as we are not delusional as regards our grasp of mathematical truths, just so long as we do have the intuitions that we think we have, then we are not machines. If indeed we truly have the intuitions that we do, then it is impossible for us to formalize (or mechanize) all of our mathematical intuitions, which means that we truly are not machines. Of course there is no *proof* that we know all that we think we know, since all that we think we know can't be formalized; that, after all, is incompleteness. This is why we can't rigorously prove that we're not machines. The incompleteness theorem, by showing the limits of formalization, both suggests that our minds transcend machines and makes it impossible to *prove* that our minds transcend machines. Again, an almost-paradox.

So Gödel was cautious about the consequences for human

nature of his incompleteness theorem. Though he did have intuitions concerning the nature of the mind, he did not, scrupulous logician that he was, deduce any such conclusions from his incompleteness theorems alone. For Gödel, the distinction between intuitions and rigorous proof was always vividly clear. After all, it was the unavoidability of that very distinction that had been so strongly suggested by his famous proof.

The second disjunct in Gödel's disjunctive conclusion concerning our mathematically knowing minds, then, consists in this possibility: we are delusional in our claims to a mathematical knowledge that exceeds all formalization. This possibility—its being precisely the possibility that gave Gödel pause—is particularly interesting when we consider an aspect of Gödel's opaque inner life that we have touched upon before: his own serious delusions.

Gödel's theorems are darkly mirrored in the predicament of psychopathology: Just as no proof of the consistency of a formal system can be accomplished within the system itself, so, too, no validation of our rationality—of our very sanity—can be accomplished using our rationality itself. How can a person, operating within a system of beliefs, including beliefs about beliefs, get outside that system to determine whether it is rational? If your entire system becomes infected with madness, including the very rules by which you reason, then how can you ever reason your way out of your madness?[11]

11 The dark mirroring is reflected in a remark of Furtwängler's, who had been Gödel's favorite mathematics professor at university, reported by Olga Taussky-Todd in her memoir of Gödel. "Auguste Bick has supplied me with an amusing remark by Furtwängler concerning Gödel's result, when the latter had one of his paranoia attacks: 'Is his illness a consequence of proving the nonprovability or is his illness necessary for such an occupation?' "

As one textbook on psychopathology puts it: "Delusions may be *systematized* into highly developed and rationalized schemes which have a high degree of internal consistency once the basic premise is granted. . . . The delusion frequently may appear logical, although exceedingly intricate and complex."

Paranoia isn't the abandonment of rationality. Rather, it is rationality run amuck, the inventive search for explanations turned relentless. A psychologist friend of mine put it this way: "A paranoid person is irrationally rational. . . . Paranoid thinking is characterized not by illogic, but by a misguided logic, by logic run wild."

It's ironic to conclude this very chapter, conveying some small sense of the superhuman beauty of Gödel's incomparable proof, with remarks on the tragic parallel between the limitations of proof ingeniously demonstrated by Aristotle's successor and the predicament of psychopathology.

IV
Gödel's Incompleteness

Pink Flamingo

There it was, inconceivably, *K. Goedel*, listed just like any other name in the bright orange Princeton community phonebook.

It was a sweetly surreal moment. I had just arrived as a graduate student at Princeton and, just for the thrilling improbability of it, had looked up the name of the town's most dazzling mind, the reigning, if reclusive, god at the fabled Institute for Advanced Study, a bucolic three-minute stroll from where I was living.

It seemed almost axiomatic to me in those days that the greatest *mathematical* mind, which happened at that moment in history to be identical with Kurt Gödel's mind, was necessarily identical with the greatest of *all* minds. *K. Goedel.* It was like opening up the local phonebook and finding *B. Spinoza* or *I. Newton.*

The Princeton community phonebook had offered me the unbelievable largesse not only of a telephone number but also of a street address for Gödel. Of course, once I had this information there was nothing else for me to do but to hop on my bike and pedal my way over to see the house on 145 Linden Lane. It was a simple wooden affair, orthogonally situated in relation to the street, unlike all the other forward-facing houses, and this singularity seemed somehow just right. The place itself was compact and modest, vaguely "European" with its red-tiled roof. By comparison, the house at 112 Mercer Street where Einstein had lived had been a mansion (and it wasn't).

The neighborhood certainly wasn't Princeton's choicest. It was a hot September day and the street was treeless, depressingly exposed to the high noon sun. There wasn't a soul stirring around the Gödel house, but the visit did manage to deliver yet another surreal wallop. The sliver of a front yard was completely dominated by one of those pink, plastic flamingoes that stands poised on one skinny leg.

I stared in disbelief at the bird. How could a man who had produced one of the most exquisite masterpieces of human thought have planted a pink flamingo on his front lawn?

Of course, there was a Mrs. Gödel, a former cabaret dancer, by popular report. Visions of *Der Blaue Engel* danced through my head, and I hastily attributed the lawn ornament to a Marlene Dietrich–type turned unlikely New Jersey Hausfrau.

I was hardly the only Princeton denizen who was fascinated with the elusive celebrity of pure thought in our midst. I once found the philosopher Richard Rorty standing in a bit of a daze in Davidson's food market. He told me in hushed tones that he'd just seen Gödel in the frozen food aisle, pushing his food cart. I went tearing through the aisles, but the phantom of logic had vanished.

"What was he buying?" I asked Rorty, for the rumor was that the man ate next to nothing. Rorty shook his head lugubriously and said he'd been too stunned to notice.

"But I guess we can assume it was something frozen."

I remember more than one party in which we, graduate students and faculty members alike—philosophers, mathematicians, physicists—sat in a circle and traded stories of Gödel. Someone had noticed that every book related to Leibniz in Firestone Library had been checked out to a *K. Goedel.* The library slips quickly disappeared, the lucky ones who got there first carrying them off as trophies.

At one party, a fellow graduate student alleged that someone had snuck up on Gödel as he sat reading in his office, to peer over his shoulder at the book, which was (uncorroborated) Ovid's love poetry in the original Latin. This same graduate student (now a prominent philosopher who shall go unnamed) at one party in which things got a bit out of hand, actually called Gödel at home, when the question arose as to whether the international phone system could become sufficiently complex to become conscious. I think I remember that he slammed down the phone when he heard Mrs. Gödel call out "Kurtsy!"

We all speculated about what our hero might be working on. There were strange rumors about a proof for God's existence—which turned out to be veridical. Gödel, like Leibniz, believed that some version of the infamous "ontological proof for God's existence" was valid. This is an argument that tries to deduce the existence of God from the right definition of God.[1]

1 The earliest version of the ontological argument was St. Anselm's, and it goes something like this: God is, by definition, that than which nothing

He mentioned to at least one colleague at the Institute, the philosopher Morton White, that there remained just one step to be clinched before his new version of the ontological argument for God's existence could be published.

The tone of our fascination with Gödel wasn't consistently reverential. There was even a decided undercurrent of flippancy. We found it hilarious, for example, that the greatest logician since Aristotle deluded himself into believing that God's existence could be proved a priori, that he was perhaps contemplating the day when atheists would be brought round by a good stiff course in quantificational logic. The stories we swiped were gratifying in the way that stories of geniuses acting oddly are always gratifying; we cut our prodigies down to human size, domesticate their grandeur into cuddliness with tales of their quotidian weirdness. Sometimes we miss seeing what is truly human within these prodigious talents, yet one more irony that Gödel's story suggests.

Gödel's story has forced me to confront a more personal irony, too, since it has required me to reacquaint myself with a field, mathematical logic, in which I had once been so much more deeply immersed and fluent. Years of a different kind of immersion, in the shadowy realm of fiction, interceded. It's not that fiction forsakes hard logic, though fiction's is not logic as we formally understand it; and it's not that fiction's logic isn't as elusive, complex, and startling as mathematical logic. Still, the logic of fiction is something quite different

---

greater can be conceived. God, therefore, cannot be conceived of as not existing, for otherwise we can conceive of Him as being greater, viz. by existing. It is therefore inconceivable that God not exist; ergo He exists.'

from formal proofs, and I know that I was the better mathematician in my youth, when I sat with others and callowly swapped stories of Gödel's genius and looniness. Yet I wonder whether I could have understood his story in quite the same way, back in the days when I was more blinded by lucidity.

Gödel, recluse though he was, made a rather surprise appearance once at an Institute garden party held for new temporary members in 1973, and, as Oskar Morgenstern wrote in his journal, the logician was in especially droll form that evening, ending up "holding court in the midst of a group of young logicians." I was there in that group, one of the acolytes agape at the god. There was a giant tent spread on the lawn behind Olden Farm, the domicile of the president of the Institute, who was then Karl Kaysen. It was a balmy October afternoon and Gödel, dapper in a dark suit, was also muffled in a long woolen scarf. I read somewhere that his height was 5' 6", but he seemed even smaller to me, and of course he was bird-thin. We all knew that the man barely ate. He was, as Morgenstern described it, in rare form (only I did not yet know how truly rare it was), clearly trying to make the youngsters feel welcome. We were mostly awed into stupidity (certainly I was). So we didn't ply him with the questions that we all wished we had, as we commiserated with one another after he had, with a brief nod and good wishes for our future work, disappeared into the falling dusk. I remember particularly regretting that I had not gotten up the nerve to ask him what he thought of the paper that the Oxford philosopher John Lucas had published, claiming that conclusions in the philosophy of mind followed from the first incompleteness theorem.

We all agreed that we wished we had asked him about what

he was working on now. He was said to go almost every day to his office at the Institute, to sit quietly there and work. What conceptual revolutions had the gnomish logician laid in wait for us? Though the number of his publications didn't amount to much—the sum total of pages equalling less than 100—in content each one had been far more than merely remarkable. But the last time he had published had been in 1958: a consistency proof for arithmetic in the journal *Dialectica.*[2]

This particular issue of the journal was a *Festschrift* in honor of the seventieth birthday of the mathematician Paul Bernays, the former Hilbert assistant who had not found, since having been dismissed as a non-Aryan from Göttingen, any academic post commensurate with his ability. It was Paul Bernays who had first presented a fully worked-out proof of the second incompleteness theorem, having learned the details on board the SS *Georgia* from his shipmate, Gödel. And Bernays had improved on von Neumann's axiomatization of set theory in a way of which Gödel highly approved and had used in his own work on set theory. So it was altogether fitting that Gödel should overcome his reluctance to publish in order to contribute something to the *Festschrift.*

Gödel had written about a new sort of proof for the consistency of arithmetic, one which was not finitary, and so therefore was consistent with his second incompleteness theorem. (But since it wasn't finitary it didn't meet the requirements of Hilbert's challenge.) He had lectured on this new sort of proof for consistency at Yale and at the Institute back in 1941. The 1958 paper was a beautifully concise statement of those ideas. Still the article hadn't contained any new results.

2 This was in fact the last thing he published in his lifetime.

People reported that there were notebooks upon notebooks of ideas that he had never published.[3] And why was he so reluctant to publish? There is much evidence that Gödel suspected that his ideas would be greeted skeptically and dismissively. Hao Wang wrote: "Gödel would probably have published more if he found himself living in a more sympathetic philosophical community. For instance, he declined to speak to what he expected to be a hostile audience."

More and more reclusive, once he lost the great friend of his life in Einstein, he was not inclined to air his views in a climate he judged to be perhaps as positivist as what he'd known back in the grungy room at the University of Vienna where the Schlick group had met on Thursday evenings at six. His sense of a world increasingly under the sway of the long-scattered Vienna Circle wasn't entirely unfounded. As Feigl recounts in his essay "The Wiener Kreis in America," positivists such as

3 In the *Nachlass* there's a sheet of paper on which Gödel had listed, probably in 1970 according to Hao Wang, all his unpublished work, from 1940 on. It reads something like:

1. About one thousand 6 x 8-inch stenographic pages of clearly written philosophical notes (= philosophical assertions).
2. Two philosophical papers almost ready for print [His paper on relativity and Kant's philosophy and his paper on syntax and mathematics, originally intended for the Carnap *Festschrift*, but never published by him.]
3. Several thousand pages of philosophical excerpts and [notes on the] literature.
4. The clearly written *proofs of* my [his] cosmological results.
5. About six hundred clearly written pages of set theoretical and logical results, questions and conjectures (to some extent *outstripped* by recent developments).
6. Many notes on intuitionism and other foundational questions.

himself and men like Hans Reichenbach and Peter Hempel, who had come to America to flee the Nazis, had had much success in bruiting their ideas abroad. In England there was the highly influential A. J. Ayer, whose *Language, Truth, and Logic* had largely been constructed out of what he'd heard in Vienna. Harvard's Willard Van Orman Quine, who had also visited with the Vienna Circle and imbibed their general outlook (though he was to disagree with them on specifics in such articles as "Two Dogmas of Empiricism" in his *From A Logical Point of View*) became the most dominant force in American philosophy. Wittgenstein's name posthumously loomed ever more prominently, the awed inclination to accept him a priori (prior even to understanding what he might have meant) persisting in analytic circles, even in the absence of his persuasive presence. And in physics departments the positivistic outlook of Niels Bohr and Werner Heisenberg had pretty much become the party line, where it still waxes strong. (It might be an interesting study to compare the figures of Niels Bohr and Ludwig Wittgenstein, both of them as charismatic as they were obscure, their obscurities pointing toward the same sort of conclusion: a *prohibition* against asking the sorts of questions that seek to make a connection between the abstract thought of their respective disciplines and objective reality.)

So Gödel wasn't being particularly paranoid in judging the climate of ideas as inimical to his own, though he perhaps both overestimated the degree of "positivism" in American universities and also underestimated his own reputation within the community and the commensurate respect that would have been accorded his views, the degree to which he might perhaps have even influenced the prevalent ideology had he braced himself to enter the fray. But that had never been his way. He certainly had never been afraid to privately dissent from the dominant views of his day; but there was a rigid reluctance to publicly voice his adversarial position in any terms other than conclusive proof.

I mentioned to the philosopher Morton White that I had come, in writing this book, to regard Gödel as an intellectual exile, or, at least, as someone who had felt himself to be in exile. White thought for several moments and came up with this story. When he had been on the faculty at Harvard, he had been instrumental in having Gödel invited to deliver the prestigious William James series of lectures and had been chagrined when Gödel declined. This would have been in the 1960s. When White himself came to the Institute as a permanent member in 1970, he remembers having asked Gödel why he had turned down the invitation. Gödel's answer had come in two parts.

First of all, he'd said, the Harvard department was too "empiricist," and he thought they'd have been critical of what he had to say. Second of all—and this part of the answer, White told me, had really interested him—Gödel felt he would have been doing an injustice to the ideas themselves, because he hadn't yet completed them; to expose them prematurely to an unsympathetic audience would be acting unjustly toward them.

So it seems, at least from this story, that his reluctance to voice his unfashionable intuitions in any form that fell short of a proof was not only a matter of his own distaste for intellectual wrangling but also connected with a perceived ethical obligation toward the ideas themselves, which is appropriate for an impassioned Platonist.

In 1964, Paul Benacerraf, of Princeton University's philosophy department, and Hilary Putnam, of Harvard's, edited a book entitled *Philosophy of Mathematics* and they wanted permission to include two of Gödel's articles, "Russell's Mathematical Logic" and "What is Cantor's Continuum Problem?" In the latter essay, revised and expanded for this volume, Gödel allows himself to state in clear and distinct terms the metaphysical Platonism to which he had subscribed, even as he'd sat, a backbencher among the positivists, quietly listening to the members of the Vienna Circle proclaiming the everlasting end of metaphysics, that is, of any assertions of existence that go beyond the empirically verifiable:

> [T]he objects of transfinite set theory . . . clearly do not belong to the physical world and even their indirect connection with physical experience is very loose. . . .
>
> But despite their remoteness from sense experience, we do have something like a perception also of the objects of set theory, as is seen from the fact that axioms force themselves upon us as being true. I don't see any reason why we should have less confidence in this kind of perception, i.e. in mathematical intuition, than in sense perception, which induces us to build up physical theories and to expect that future sense perceptions will agree with them and, moreover, to believe that a question not decidable now has meaning and may be decided in the future. The set-theoretical paradoxes are hardly any more troublesome for mathematics than deceptions of the senses are for physics.

Gödel explains in the article how Cantor's continuum hypothesis has been shown to be independent of the axioms of set theory, and his reasons for believing that the hypothesis is actually false. (His was the part of the proof [of the undecidability of the continuum hypothesis] that showed that the continuum hypothesis can't be proved false on the basis of present axioms of set theory, in other words that it's consistent with the axioms of set theory, so his believing that it's nonetheless false is particularly interesting. Paul Cohen proved that the continuum hypothesis, on the basis of the axioms of set theory, can't be proved true either. So together they proved the undecidability of the continuum hypothesis.) He connects his Platonist belief in the objective truth or falsity of such undecidable propositions as the continuum hypothesis with his own incompleteness result:

> What, however, perhaps more than anything else, justifies the acceptance of this criterion of truth in set theory is the fact that continued appeals to mathematical intuition are necessary not only for obtaining unambiguous answers to the questions of transfinite set theory, but also for the solution of the problems of finitary number theory (of the type of Goldbach's conjecture [which, you may remember, asserts that every even number larger than two is the sum of two primes]) where the meaningfulness and unambiguity of the concepts entering into them can hardly be doubted. This follows from the fact that for every axiomatic system there are infinitely many undecidable propositions of this type.

Benacerraf recounted for me how Gödel would call either him or Putnam every day to voice his ambivalence and misgivings about having his articles included in their volume, extending his permission one day only to withdraw it the next, and then rethinking the withdrawal the day after that. He was afraid that the two "positivist" editors would use their introduction to attack his ideas. Only when the two, singly and repeatedly, had promised him that they intended only to place each of the articles in its proper context and had no intention whatsoever of evaluating any of the chosen contributions did he finally agree to have his articles included.

Was there any basis, I asked Benacerraf, for Gödel's thinking that either he or Putnam was a positivist?

"Well, Putnam, at some stage at least, sure. After all, his dissertation advisor had been Reichenbach."

The Benacerraf /Putnam volume was divided into four sections: Part One: *The Foundations of Mathematics;* Part Two: *The Existence of Mathematical Objects;* Part Three: *Mathematical Truth*; and Part Four: *Wittgenstein on Mathematics.* Though the volume includes the writings of Frege, Hilbert, Gödel—all the leaders in foundations—only Wittgenstein, who had never accepted Gödel's theorems as important, is judged sufficiently significant to merit an entire section. How, one wonders, did Gödel react to his author's copy?

Gödel had clearly expected that the full implications of his theorems would be as transparent to others as they had seemed to him, and he was not beyond the perfectly normal human reactions of disappointment and even resentment (though this was usually hidden behind the opacity of his thick reserve). He'd complained to Olga Taussky-Todd, as she reports in her memoir of him, that Hilbert still, even after Gödel's proof, had continued to espouse formalism. "He spoke to me about this, I think in Zurich, and lashed out against Hilbert's paper 'Tertium non datur' [Goettinger Nachr. 1932], saying something like 'how can he write such a paper after what I have done?'"

He seemed to have felt increasingly alone and embattled in

the highest turret of Reine Vernunft and took refuge in the sort of profound isolation that few spots on earth can afford with such abundant completeness—if that's what one is after—as the Institute for Advanced Study.

The Coffee Is Wretched

Gödel first came to the Institute for the academic year 1933–34, which was the first year of its operation. The mathematician Veblen, one of Flexner's first appointments to the nascent Institute, had met the young logician in Vienna and had been sufficiently impressed to bring him to New Jersey for a temporary visit. Of course, von Neumann, who was also spending at least some of his time at Princeton, was quite interested in the logician who had spoken of his revolutionary theorems in one crisp compressed sentence delivered in Königsberg. Gödel had not wanted to lecture his first semester in America, being still uncertain of his English, but by the second semester he gave a series of lectures on incompleteness. This is how it came to be that the talk among von Neumann and his circle was filled with references to Gödel when Alan Turing came to spend the academic year 1936–37 in Princeton, so that he returned to Cambridge determined to pursue Gödelian lines of reasoning, with such successful results.

Sometime in the course of this first stay, Gödel met Einstein. Einstein had already moved permanently to Princeton, since Nazified Germany was no longer a possibility for him. It was Veblen who introduced them, but the famous friendship didn't begin until several years later when Gödel himself moved permanently to Princeton.

Gödel, as a classified Aryan, went back to Vienna at the end of the academic year. Menger noted that when Gödel returned he seemed even more fragile:

> Gödel was more withdrawn after his return from America than before; but he still conversed with visitors to the Colloquium. . . . To all the members of the Colloquium Gödel was generous with opinions and advice in mathematical and logical questions. He consistently perceived problematic points quickly and thoroughly and made replies with greatest precision in a minimum of words, often opening up novel aspects for the inquirer. He expressed all this as if it were completely a matter of course, but often with a certain shyness whose charm awoke warm personal feelings for him in many a listener.

Gödel in fact spent a few weeks in a sanatorium after returning to Vienna, where the psychiatrist Julius Wagner-Jauregg, who had won a Nobel prize in 1927, diagnosed a nervous breakdown brought on by overwork. Of course, it might not have been just overwork. Gödel's crisis came soon after Moritz Schlick had been shot dead on the stairs of the university. This event was profoundly unsettling for even the most stable of people. The murder signified the effective end of the Vienna Circle, although its influence would continue to spread, especially with most of its former members soon required to find refuge outside Europe. In any case, Gödel recovered sufficiently, after a few weeks in the sanatorium, and returned to teaching his course on topics in mathematical logic.

Gödel's position at the University of Vienna was rather a lowly one. He became a *Privatdozent* in Vienna in 1933. A Privatdozent is granted the right to lecture, though he receives no salary. For the honor, a candidate has to write a second dissertation. Gödel's proof of the completeness of the predicate calculus (limpid logic) had constituted his Ph.D. dissertation. His proof of the incompleteness of arithmetic was submitted for his second dissertation, the *Habilitationsschrift*. The commission to consider Gödel's application met on 25 November 1932. A candidate for the *Dozentur* requires a sponsor, and Hans Hahn, his dissertation advisor, served as Gödel's sponsor. Hahn testified to the committee that Gödel's dissertation was of great scientific worth and that the *Habilitationsschrift* was "an achievement of the first rank" that had "attracted the greatest attention in scientific circles" and was already destined to go down in the history of mathematics (which seems, on the whole, rather faint praise).[4] Hahn made the not terribly audacious judgment that Gödel's work far exceeded the requirements for the *Habilitation* and the committee unanimously agreed.

However, this wasn't the last hoop a candidate had to jump through before being declared a Privatdozent. The entire faculty had to take a vote, not only on the candidate's scientific worth but also on his personal worthiness as well. "The results, as recorded by the dean in his report of 17 February 1933 to the ministry of instruction were, on the question of his character, fifty-one in the affirmative and one 'no.' On the question of his

4 *Time* magazine, in commemoration of the end of the last millennium, devoted a few special issues to the 100 greatest minds of the last century. Kurt Gödel was cited as the century's greatest mathematician. Interestingly, Ludwig Wittgenstein and Alan Turing also made the list, and Albert Einstein was chosen as the greatest mind of the century.

scientific merit, there were forty nine 'yes's and one quite astounding 'no.' " (Two professors must have left between the two vote-takings.) John Dawson writes that in a private communication, "Dr. Werner Schimanovich has reported that the naysayer was Professor Wirtinger, who thought that the incompleteness paper overlapped too much with the dissertation," an almost inconceivably wrongheaded appraisal, since the dissertation had proved the completeness, not incompleteness, of a formal system, viz. that of the predicate calculus (or limpid logic).[5] But that there should have been any dissent at all on the question of whether Gödel's incompleteness theorems were worthy of earning him a place on the lowest rung of the University of Vienna is bizarre. What was Gödel's inner response, one wonders, to the less-than-unanimous vote? Disdain, resentment, an increased sense of insecurity? Gödel, of course, knew the significance of his theorems in all their mathematical and metamathematical splendor. But his intellectual audacity was so strangely coupled with diffidence that, again, it would be foolhardy to try to guess the affect behind the opacity.

Gödel was never to learn the most elementary of lessons with regard to maneuvering for position and status. One can well imagine that any Viennese professor trying to judge the importance of the mathematical work from the self-importance of the mathematician would have been misled. It was to be no different at the Institute. It was only in 1953—after he had

5 Wirtinger was a mathematician who had reportedly become embittered and withdrawn after his colleague, Professor Furtwängler, got a prize for an important result in algebraic number theory. What a way to earn a footnote in the story of the most important mathematical result of the twentieth century.

received an honorary degree from Harvard, which cited the incompleteness theorems as the most important mathematical discovery of the past century, and had also been elected to the National Academy of Sciences—that he was, at long last, made a permanent member of the Institute. Once again, the impetus came from von Neumann, who is reported to have said, "How can any of us be called professor when Gödel is not?" Given his lack of worldly status, it is perhaps no wonder that his wife always considered his older brother the more successful of the two, since he was a *medical* doctor.

Gödel had married Adele Nimbursky (nee Porkert) in 1938. Hao Wang, who in the spring of 1976 (the year of Gödel's death) had decided to write a précis of Gödel's intellectual development, giving the logician the opportunity to comment on it, writes (in an endnote): "G married Adele Porkert on 20 September 1938. He asked me to delete this information from an early draft on the ground that his wife has no direct influence on his work." The Gödel marriage was, according to just about everyone, weird. Gödel's mother, in particular, found her son's matrimonial choice inexplicable. His father had already died; but then he had never been close to his father. His relationship with his mother, however, was entirely different, so his unexpected choice in a bride was some cause for vexation.

The catalogue of maternal complaints against Adele were primarily these: she was a divorced woman of the wrong religion (Catholic), wrong class (lower), wrong age (six years older than Gödel), wrong appearance (she had a port-colored stain on the side of her face), and, perhaps most seriously, wrong occupation (she'd been a cabaret dancer; she would tell people it had been ballet, but it hadn't).

Princeton's "society" for the most part concurred with Gödel's mother. When Adele arrived in Princeton, Oskar Morgenstern described her as a typical Viennese washerwoman and correctly predicted that she would be a dismal social failure. "Of course Gödel himself is half crazy," he recorded in his journal. The verification of Morgenstern's prediction caused Adele Gödel much grief, though of course her husband could not have cared less. Gödel was as indifferent to the snubs as he was indifferent to the causes for them. He had married a Hausfrau-caretaker and Adele proved as capable of caring for his frail body and soul as could be expected of a mere mortal. Adele told a neighbor friend that even back in Vienna, when she first became involved with Gödel, she used to taste his food for him, foreshadowing the dark sieges of paranoia that would increasingly seize hold of him.

Soon after his marriage in Vienna, Gödel set sail, without his bride, once more for Princeton, where von Neumann had drummed up great interest in Gödel's theorems. He lectured that semester at the Institute on his discoveries in set theory (concerning both the axioms of choice[6] and the continuum

6 The axiom of choice is concerned with collections of sets, particularly infinite collections. There are various ways of stating the axiom. In fact there's a whole book, by H. Rubin and J. Rubin, entitled *Equivalents of the Axiom of Choice.* A simple version of the axiom is: For any set of non-empty disjoint sets (sets that have no members in common), there exists a set consisting of exactly one member of each of the non-empty sets. In other words, if you have a bunch of sets that don't overlap with each other, then, roughly speaking, you can form a set by choosing one member of each set in the bunch. (You really need the axiom only when the bunch is infinite.) Another way of stating the axiom of choice is: For any set of non-empty sets, there exists a function that assigns to each one of these non-

hypothesis). After spending the autumn term at the Institute, he went on to South Bend, Indiana, to spend the spring term at Notre Dame. His visit to Notre Dame was due to Menger, who had decided to emigrate and had ended up in Indiana.

While Gödel was in South Bend, Czechoslovakia was handed over to Hitler. This was the fateful year of 1939, the entire world holding its breath. To the dismay of Menger, Gödel, instead of sending for Adele and making plans to

---

empty sets one of its members. Infinitely many *choices* (hence the name of the axiom) may be required, which is why the axiom has received so much attention. The axiom is saying that a certain set exists, even though the set is not really specified or constructed. The axiom of choice is probably the second most discussed axiom of mathematics, right after Euclid's parallels postulate. Like the parallels postulate, the axiom of choice was proved to be independent of the other axioms, in this case of set theory. Gödel proved the first part of the independence by showing that the axiom is consistent with the other axioms of set theory; and then (as with the continuum hypothesis) Paul Cohen completed the proof (in 1963) by showing that the negation of the axiom of choice is consistent with the other axioms. Just as the proof of the logical independence of the parallels postulate gave rise to non-Euclidean geometry, so, too, there's a non-Cantorian form of set theory that uses the *negation* of the axiom of choice. But though the very idea of the infinite number of choices involved in the axiom of choice might make mathematicians a bit queasy, most mathematicians don't hesitate to avail themselves of the axiom in constructing their (nonconstructive) proofs, because it has so many important applications in practically all branches of mathematics that its rejection would seriously manacle mathematicians. It's not clear when Gödel began to think about set theory, and it's not clear when he proved that the axiom of choice is consistent with the other axioms of set theory. He didn't tell anyone about his proof until the following year when he was back again in Princeton. Not surprisingly, it was von Neumann who received the confidence of the important new result, which Gödel published in 1938.

resettle in America, insisted on returning to Vienna. He was extremely incensed that his *Dozentur* was in danger of being revoked by the New Order, and felt that he must hurry back and see that his rights weren't violated. Menger, devoted to Gödel, found his fondness severely tried.

> He had complained about the revocation of his dozentship and had spoken about violated rights. "How can one speak of rights in the present situation?" I asked. "And what practical value can even *rights* at the University of Vienna have for you under such circumstances?" But despite pleas and warnings by all his acquaintances at Notre Dame and Princeton, he was determined to go to Vienna; and he went.

A man who could be terrorized by a refrigerator, convinced that it was emitting poisonous gases, returned to a Vienna overrun with Nazis to secure "his rights."

Gödel's world back in Vienna was now thoroughly Nazified. Menger had written to Veblen, while still in Vienna, that "whereas I . . . don't believe that Austria has more than 45% Nazis, the percentage at the universities is certainly 75% and among the mathematicians I have to do with . . . [apart from] some pupils of mine, not far from 100%."

Gödel was decidedly not anti-Semitic. He never took the slightest offense when others assumed he was Jewish; he simply corrected the error "for the sake of the truth," as he had written in the unmailed correction to Bertrand Russell's autobiography. The group of thinkers with whom he associated in Vienna were for the most part Jewish. Gödel neatly instantiates the tongue-in-cheek advice that the satirist Leon Hirshfeld gave to travelers: "Be careful during your stay in

Vienna not to be too interesting or original, otherwise you might behind your back suddenly be called Jewish."

Though he didn't in the least partake of the crudely simplifying racial theories that were having such a popular run in Vienna and elsewhere, Gödel was dismayingly indifferent to the plight of the victims of those racial theories. The German (Jewish) positivist philosopher, Gustav Bergmann, recalled to John Dawson that shortly after arriving in America in October 1938, he was invited to lunch with Gödel, who asked, with clueless charm, "And what brings you to America, Herr Bergmann?"

For Menger, there was a limit to how much one could forgive clueless genius:

> During the summer I heard nothing from Gödel. But on August 30, 1939, one of the few days between the Hitler-Stalin pact and the entry of German troops into Poland which unleashed the second world war, he wrote me a letter that may well represent a record for unconcern on the threshold of world-shaking events: "Since the end of June I have been here in Vienna again and had a great many tasks to perform so it was unfortunately not possible to write up anything for the Colloquium. How did the examinations turn out for my logic lectures? . . . In the fall I hope to be back in Princeton."

Menger's affection for Gödel considerably cooled, not to return until decades later, at the end of Gödel's life, when he came to understand more completely the deep and abiding strangeness of the logician.

So there was Gödel in Vienna in 1939. But it was a Vienna

sadly changed, as even Gödel must have apprehended, however dimly. The old Circle was no more, of course. Schlick had been murdered by a psychotic student who was then transformed into a hero in the Nazi press; Feigl, Carnap, and Menger had fled from the increasingly poisonous atmosphere. Hans Hahn had died of cancer on 24 February 1934 at the age of 55—one day before a group of Nazis stormed the chancellery in Vienna and assassinated Dollfuss in a failed putsch.

Nonetheless Gödel apparently intended to remain in Vienna. His wife and he had signed a new lease on their apartment and had even arranged for some renovations. He had, in addition to renewing his apartment contract, made an attempt to become a *Dozent neuer Ordnung*. The authorities "of the New Order" who vetted his application noted that he was "well recommended scientifically" but that his *Habilitation* had been supervised by "the Jewish professor Hahn," and that it "redound[ed] to his discredit" that he had "always traveled in liberal-Jewish circles." To be fair to him, it was observed that "mathematics was at that time strongly *verjudet*," or Jewified. The authority in charge of Gödel's case, Dr. A. Marchet, the *Dozentbundsführer*, could find no statements by Gödel on record opposing National Socialism, but none supporting it either. Dr. Marchet's decision, concerning the status of Kurt Gödel in the New Order, was not forthcoming; he could not bring himself either to approve Gödel or, for the time being, reject him.

The event that seemed to have precipitated his decision to leave Vienna was that Gödel was once again mistaken for a Jew, this time far more threateningly, since it was by a group of young thugs in the vicinity of the university. He was roughed up, his glasses smashed on the sidewalk. Gödel didn't look *un*Jewish, especially in his habitual long overcoat, his fedora, his heavy-framed thick eyeglasses. He looked like an intellectual, and that was incriminating enough. The doughty Adele, displaying the protective skills for which he no doubt married her, fought the fascists off with her umbrella.

Gödel had also received the shock of being declared fit for military service, which no one had been expecting. (The draft board in America almost automatically classified him 4-F—disqualified for physical reasons.) The authorities had refused to excuse him from military service on the basis of the heart that Gödel continued to maintain had been damaged by his bout of rheumatic fever at age eight. They also overlooked the more convincing evidence of his mental instability, including, by this time, a few more stays in sanatoria. This is a fortunate oversight, since "mental defectives" were dispatched not to the front but to places that more efficiently eliminated them.

It was difficult to obtain a leave of absence from the University of Vienna, as well as permission to leave Austria for America from both the Austrians and the Americans, but Gödel's Princeton supporters—Abraham Flexner; Flexner's successor as director of the Institute,[7] Frank Aydelotte; Veblen; and von Neumann—all joined forces to make it possible for Gödel to cross the Atlantic. Gödel credited, in particular, Aydelotte's letter to the chargé d'affaires at the German Embassy in Washington. In that letter, dated 1 December 1939, the new director of the Institute testified that Gödel was an Aryan who was one of the greatest mathematicians in the

7 Flexner's resignation was the result of his having made two appointments, both of economists, without first consulting the touchy faculty. The mathematicians were particularly sensitive on this matter, foreshadowing events that were to cast a pall over Gödel's last years at the Institute.

world. "His case could hardly create a precedent," Aydelotte reasoned, "because there are so few men in the world of his scientific eminence."

The Gödels began their journey to the New World, going first through Russia and onward to Japan on the trans-Siberian railway. The German certificate of exit had required this route, and in addition, as Gödel wrote in a letter to Aydelotte, "I am told in all steamship bureaux that the danger for German citizens to be arrested by the English is very great on the Atlantic." They arrived in Yokohama on 2 February 1940, a day after the ship that had been booked for them by the Institute had sailed. They had to wait for the next ship, the *President Cleveland*, which docked in San Francisco on 4 March. From there it was a matter of taking the transcontinental railroad to New York, and then finally traveling to their new home in Princeton.

This sounds like rather a drama, in the normal sense of the word. But Gödel was coolly detached from the sort of drama that escaping Nazi Europe afforded. "Gödel has come from Vienna," wrote Oskar Morgenstern in his journal. Morgenstern, too, was originally from Vienna and was naturally eager to get news of his beleaguered city from the newly arrived logician. "In his mix of profundity and otherworldliness he is very droll. . . .When questioned about Vienna, he replied 'The coffee is wretched.'"

Adele would make the return trip several times after the war to visit her mother back home, but Gödel never set foot on European soil again. His mother would have to come to Princeton, as she did several times, if they were to see each other again. In fact, so unlike the typical peripatetic academic, he barely ever strayed, for the next 38 years of his life, out of the township of Princeton.

Gödel could not even be induced to make the easily walkable trip to Princeton University when, in 1975, it finally got around to offering him an honorary doctorate. He already had such degrees from Harvard, Yale, Amherst, and Rockefeller. It had been primarily through the efforts of Paul Benacerraf that the university decided to acknowledge the genius who had become, for those of us who cared, something like the Greta Garbo of the intellectual world, wanting to be alone. However, as commencement day approached, Gödel's initial pleasure gave way to his far more characteristic tergiversation, continuing until the very morning of the event. Both Benacerraf and Simon Kochen offered to chauffeur him to the ceremony and attend to all other concerns, but Gödel ended up sitting the honor out. Perhaps he was miffed that the honor had come so late. "Ten years ago," he told Morgenstern, "such a thing . . . would have been proper." The one condition of receiving the honorary doctorate is that one show up. Therefore, although Gödel was listed in the program as having received a Doctor of Science, the program lied. Here, nevertheless, is the lovely citation: "His revolutionary analysis of received methods of proof in that most familiar and elementary branch of mathematics, the arithmetic of whole numbers, has shaken the foundations of our understanding both of the human mind and the scope of one of its favorite instruments—the axiomatic method. Like all important revolutions, his has not only shown the limits of old methods, but also has proved a fertile source of fresh ones, leaving new and flourishing disciplines in its wake. Logic, mathematics, and philosophy all continue to gain immeasurably from his genius."

One anecdote of Gödel in America bears repeating, and this concerns his becoming an American citizen. This is perhaps the most famous story told about Gödel. (It comes to us by way of Morgenstern.) Not only does it involve Einstein playing straight man to that wild guy, Gödel, but it also sets off the cuddly eccentricity of the genius, and everybody seems to enjoy these kinds of tales. (At a reading I gave at my college of a chapter of this book, the "question period" quickly degenerated into a session of trading such stories about Gödel. Unfortunately, I'd already heard them all.)

Gödel had taken the whole matter of American citizenship very seriously, studying thoroughly in preparation for his exam; so thoroughly, in fact, that he made, he believed, a disturbing discovery: there is an internal contradiction in the American Constitution that would allow its democracy to deteriorate into tyranny.[8]

In a state of high consternation Gödel revealed his finding to Morgenstern. There was always a strongly legalistic bent to Gödel, a fascination with examining the meaning and implications of man-made laws that faintly mirrored his interest in the eternal laws of logic. The economist was

8 Unfortunately, Morgenstern's account, and so all the others that derive from it, omits mention of the precise constitutional flaw. I asked John Dawson whether he knew what it was supposed to be, and he e-mailed back: "No, I don't, though many have asked that question. There is a set of shorthand notes in Gödel's *Nachlass* concerned with American government (presumably made while he was studying for the citizenship examination) that *might* contain the answer, but transcribing that particular item has never had as much priority as the mathematical material" (3 January 2004).

both amused by Gödel's argument and concerned, because he knew that, Gödel being Gödel, he might very well behave in such a way as to jeopardize his eagerly anticipated citizenship. Morgenstern consulted Einstein on how best to handle the logician.

On the day of Gödel's citizenship test, 5 December 1947, Morgenstern and Einstein arrived to take Gödel to Trenton. Morgenstern was the designated driver and Einstein the designated distracter. As soon as Gödel stepped into the car, Einstein, not giving him a chance to speak, greeted him with a diverting joke.

"Well, are you ready for your next-to-last test?'"

"What do you mean, 'next to last'?"

"Very simple. The last will be when you step into your grave." Old-world hilarity.

Einstein continued on, telling story after story, including one about a recent autograph hound. He observed that such people are the last of the cannibals, in that they seek to take possession of the souls of those they ingest. And so the three members of the Institute for Advanced Study managed to arrive at the Trenton federal courthouse. There were several applicants ahead of them and so Einstein was resigned to keeping up his diversionary shtick; but fortunately it turned out that the judge, whose name was Philip Forman, was the very one who had administered the oath of citizenship to Einstein some years before and he ushered the three men into his chambers immediately.

Einstein and Forman chatted for a while and Gödel, sitting quietly and biding his time, seemed all but forgotten. Eventually, though, Forman got on with the business of the day.

"Up to now you have held German citizenship."

Immediately, Gödel corrected the judicial error: *Austrian* citizenship.

Duly corrected, the judge continued.

"In any case, it was under an evil dictatorship. Fortunately, that is not possible in America."

This was just the opening the logician had been waiting for.

"On the contrary," he objected, "I know precisely how it can happen here," and he began to launch into his account of the flawed Constitution. Forman, Morgenstern, and Einstein exchanged meaningful glances and the judge called a halt to Gödel's exposition, with a hasty, "You needn't go into all that," and steered the conversation round to less dangerous subjects. A few weeks after he'd taken the oath, Gödel aptly described Forman in a letter to his mother as "a very sympathetic person."

"The Logic, It Was Impossible"

Every day, the two of them, Einstein and Gödel, would walk home together from the Institute, deep in conversation, and others watched them and wondered. Gödel took great pleasure—perhaps even pride—in the friendship; the references to Einstein in his letters to his mother bear witness to this. "I keep on wondering over Einstein's walking to the Institute in such weather. But he appears to be in this respect a match of you in unreasonableness," Gödel wrote his mother teasingly on an inclement 17 February 1948. And on July 12 of that same year: "I see Einstein almost daily. He is very robust for his age. One does not see that he is already nearly seventy and he now appears also to feel completely well in terms of his health."

But it was only a few months after this, in the autumn of 1948, that Einstein, suffering attacks of pain in his upper abdomen, entered Jewish Hospital in Brooklyn for an exploratory laparotomy. An abdominal aneurysm was discovered. A few years later Einstein learned that the aneurysm was growing and, reported Helen Dukas, "We around him knew of the sword of Damocles hanging over us. He knew it, too, and waited for it, calmly and smilingly."

Einstein was careful that Gödel, so obsessed with his own health, never knew. Einstein was always extremely protective of his delicate younger friend. So Gödel wrote repeatedly to his mother of Einstein's robust health (compared to his own health problems, both real and imagined) until the letter of 25 April 1955. Einstein had died on 18 April 1955:

> The death of Einstein was of course a great shock to me, since I had not expected it at all. Exactly in the last weeks Einstein gave the impression of being completely robust. When he walked with me for half an hour to the Institute while conversing at the same time, he showed no signs of fatigue, as had been the case on many earlier occasions. Certainly I have purely personally lost very much through his death, especially since in his last days he became even nicer to me than he had been all along, and I had the feeling that he wished to be more outgoing than before. He had admittedly kept pretty much to himself with respect to personal questions. Naturally my state of health turned worse again during last week, especially in regard to sleep and appetite. But I took a strong sleeping remedy a couple of times and am now somewhat more under control again.

After Einstein's death, Gödel's sense of exile must have deepened enormously. When Einstein had been ordered by his doctor to take a rest cure, there had been nobody, as Gödel complained to his mother, for him to speak to. Now there would permanently be nobody.

His profound isolation wasn't only a matter of his intellectual estrangement from the philosophical positivism that he felt had trailed him to the New World from Vienna (which in some sense it had). On a personal level, as well, Gödel was quite completely alienated from his mathematical colleagues at the Institute. Unlike Einstein, they weren't amused by his "strange axiom," his version of Leibniz's principle of sufficient reason, which disposed him to believe that everything that happens has a thoroughly logical explanation—especially since Gödel's application of his axiom led him to believe that those in authority are indeed in authority for a sufficiently good reason. Gödel's axiom inclined him to give the Powers That Be the benefit of the doubt—they must have their good reasons for their decisions and actions, even if all empirical evidence seems to indicate that they don't—and such reasoning served to deeply divide the logician from his colleagues at the Institute.

So far as mathematics went, the mathematicians found that Gödel, logician though he was, was a ready participant in their theoretical discussions. "In fact, he knew more mathematics than I had suspected," Borel explained to me when I visited him at the Institute. "He could participate in discussion not just about logic. In mathematics, he was really a participant to our discussions."

It was in the more practical sphere that Gödel alienated his fellow mathematicians, at least as the practical presents itself at the Institute: the all-important issue of appointments. It's particularly the matter of the permanent membership that snaps the most otherworldly of Institute thinkers to full attention: Who is worthy to enter the empyrean reaches of pure reason?

Flexner had chosen mathematics, "the severest of disciplines," as his model; not only are mathematical results certain, but the relative depth and importance of the results are also certain. So, too, is the relative depth and importance of the mathematicians themselves commensurately certain. Mathematicians know exactly who among them is the best, and the best is what the Institute is all about. The Institute's mathematicians have a tradition of judging other disciplines, well, severely.

The mathematicians balked at Flexner's attempt to diversify the population at the Institute, to include scholars of economics, politics, and the humanities. Flexner managed to set up two new schools, one of economics and politics, another of humanistic studies, but the battle had been quite acrimonious and when Flexner retired four years later he was a very tired man. Getting the start-up money out of the Bamberger/Fulds—who, after all, had respect for him and his opinions—had been a piece of cake compared to getting his proposals past the mathematicians. His successor, Frank Aydelotte met, for the most part, with the mathematicians' approval. He was, in Einstein's words, "a quiet man who will not disturb people who are trying to think."

When Aydelotte retired in 1947 the directorship passed to J. Robert Oppenheimer, who had returned to teaching at Berkeley and Caltech after the successful completion of the "Manhattan project"—the war-time mission that had brought together many of America's leading physicists to develop the first atomic bomb. Oppenheimer had hesitated, as Einstein had, in making the move to Princeton. After visiting the

Institute, he spoke mockingly of its "solipsistic luminaries." Still he came, and it wasn't long before he and the mathematicians were sniping at one another.

Oppenheimer, quite understandably, was interested in strengthening the Institute's school of physics. Individual appointments brought mathematicians like von Neumann and Deane Montgomery into sharp opposition with the director. In Oppenheimer's day the entire faculty voted on every appointment. Nobody could really judge the mathematicians' work except the mathematicians themselves, though they seemed to have no trouble passing judgment on the work of the physicists, economists, historians, and humanists (yet another perk of residence in the highest turret of Reine Vernunft). Ironically, the otherworldly mathematicians were *the* force with which to be reckoned when it came to the most central practical concern at the Institute. Some have theorized, somewhat facetiously, that the trouble with the mathematicians is not just the lofty standards that they're used to employing, but that they also tend to work fewer hours of the day than other people. This leaves them ample time for mischief.

But it wasn't simply Oppenheimer's advocacy of nonmathematicians that poured fuel on the mathematicians' ire. It was, in fact, the candidacy of the mathematician John Milnor that lit the match. John Milnor was then a mathematician at Princeton University. When he had been an 18-year-old freshman at Princeton he'd heard about a conjecture of the Polish topologist, Karol Borsuk, concerning the total curvature of a knotted curve in space. Milnor figured out a proof of the conjecture and took it to his professor, saying, "I can't seem to find anything wrong with this, can you?" The professor couldn't and neither could his colleagues. A year later, Milnor had worked out a general theory of the curvature of knotted curves that had the proof of Borsuk's conjecture as a mere by-product. He'd gone on to a brilliant career, and the Institute mathematicians wanted him as one of their own. Oppenheimer opposed them, saying that there had been a pledge given to the university that the Institute would not come courting in its own backyard. The mathematicians countered that there had never been such a pledge, that Oppenheimer was constructing it ad hoc for his own ulterior motives. (The mathematicians' doubts about Oppenheimer's good faith were so deep and abiding that I could catch the echoes of them even today. One mathematician, describing the path that used to lead directly from Fuld Hall to Olden Farm, the director's residence—the very path that Einstein and Gödel would daily walk—told me that after Einstein's death Oppenheimer "for some reason suppressed the path," that is, let the grass grow over it. "I have no idea why," the mathematician concluded, giving me a dark look strongly suggestive of the sinister intent behind this path suppression.)

Gödel, as Borel explained to me, had wanted Milnor to come to the Institute as much as any of the mathematicians had. But he couldn't bring himself to oppose the director's authority. It was Gödel's unchallengeable adherence to authority's rights that provoked the rest of the mathematicians to decide "that there was just no possible argument" [with Gödel], as Borel put it to me, continuing:

> The logic was just so strange. There just could be no discussion, not even a common way to discuss the matter. He was always with the authority. Deane Montgomery and I,

> we were talking to Gödel, and the logic, it was just totally impossible. Here was a man who had had to flee the fascists in power in Austria. And yet, by his logic, one does not defy authority. The logic, it was just impossible. And it was so bad in the department that it was decided, by general consent, that from then on logic would be handled separately.

What this meant was that Gödel would no longer be included in the discussions of the mathematicians on appointment; he wasn't sent the files of prospective candidates, he wasn't solicited for his opinion, he wasn't present at the meetings. He was exiled to his own sphere: logic. He would decide on the appointments of logicians, conferring with the mathematician Hassler Whitney.

This was in 1961, and after that almost all conversation between Gödel and the other mathematicians ceased. Only Whitney kept up any contact with Gödel. And the contact was only of the most professional sort, for this is how Gödel himself wanted it. At Gödel's memorial service, Whitney recalled how he'd once gone to pay a visit to Gödel and Gödel had been taken aback, since there'd been no "precise issue" that had brought Whitney to Gödel's door.

Gödel also developed a strong preference for conducting all conversations over the telephone. Even if a colleague was a few feet's stroll away from his office at the Institute, Gödel would instruct him to use the phone.

There was a brief period, in the early seventies, when Gödel manifested what was, for him, a preternatural gregariousness. Simon Kochen told me that during this period Gödel would often call him, to catch himself up on the latest work in his field. In March of 1973, Abraham Robinson (1918–1974), a mathematician whose work Gödel admired, gave a talk at the Institute. Robinson's work had used the techniques of formal logic, many developed by Gödel in the course of his proof of the first incompleteness theorem, to solve standard problems in algebra, engendering what is called "nonstandard analysis," and such extensions of logic's reach were always encouraging to Gödel. (Simon Kochen's work as a very young logician had similarly brought formal logic to bear on a more traditional mathematical problem.) Robinson's talk prompted the usually taciturn Gödel to rise to his feet to congratulate Robinson on his work.[9] Nonstandard analysis, he said, was not "a fad of mathematical logicians" but was destined to become "the analysis of the future. . . . In coming centuries it will be considered a great oddity . . . that the first exact theory of infinitesimals was developed 300 years after the invention of differential calculus."

In the autumn of 1973, Gödel surprised everyone by holding court at that lawn party at which I got my one and only opportunity to meet him. Joseph Stalin's daughter, Svetlana Alleluyeva, a minor celebrity memoirist of the 1970s, was, I remember, at that party as well, but who had eyes for a Stalin when a Gödel was there? I never caught another glimpse of him.

Then Gödel managed to infuriate his colleagues once again, during the Institute cause celebré that produced waves so huge as to overspill onto the pages of the *NY Times* and such magazines as *Harper's* and the *Atlantic Monthly* (whose cover for the February issue of 1974 read: "Bad Days on

9 A few months after this talk, Robinson succumbed to pancreatic cancer, a death which reportedly struck Gödel hard.

Mount Olympus: The Big Shoot-Out in Princeton"). Though the director was now a different one—Karl Kaysen, who had taken over the reins from Oppenheimer in 1966[10]—the bone of contention was the same—namely, appointments.

Kaysen had come to the Institute from Harvard's economics department, but he'd taken a leave from Harvard in 1961 to work for McGeorge Bundy at the National Security Council in the Kennedy White House. This was already far too much of the real world for the mathematicians, and they—excluding Gödel—were already strongly inclined toward bristling skepticism about the new man on campus.[11] But when Kaysen proposed establishing a new school of social science, promising that he would raise all the funding for this school on his own, the mathematicians, as well as many from the School of History, girded themselves for full-scale warfare.

The first appointment was of Clifford Geertz, a cultural anthropologist from the University of Chicago, whose work addresses all aspects of culture and whose professional credentials were as unchallengable as a social scientist can hope

10 Oppenheimer had managed to merge the floundering School of Economics and Politics with the flourishing School of Humanities to create the new School of History. But by the time he stepped down—with less than six months to live—he found that half his faculty weren't speaking with him. All the mathematicians—with the exception of Gödel—were in the enemy camp. In speaking to mathematical survivors of that period even today, Oppenheimer's character is described in uncharitable terms. Memories seem never to fade at the Institute.

11 His scholarly work primarily concerned American anti-trust policy; he'd done a study of *United States v. United Shoe Machinery Corporation*. Said the mathematician André Weil, "I think he wrote his thesis about a shoe factory."

to attain. He got by. It was the candidacy of Robert Bellah, a sociologist of religion at Berkeley, that converted the hostile rumblings into pitched battle. Though of course the weapons were words, they could be quite targeted, even deadly. The mathematician André Weil,[12] for example, was quoted in the *NY Times* as saying, "Many of us started reading the worthless works of Mr. Bellah. I've seen poor candidates before, but I've never had the feeling of so utterly wasting my time." Some supporters of Bellah objected that Weil, as a mathematician employing the high standards of his discipline, would find any attempt to study religion sadly lacking in rigor. Weil countered that on the contrary, he had a personal connection with such topics; after all, his sister had been the famous mystic, Simone Weil.

Gödel once again agreed in principle with the mathematicians. Borel told me that Gödel had said that Bellah's appointment would be the weakest in the history of the Institute. At the general faculty meeting during which Bellah's candidacy was discussed—a meeting in which many unfortunate things were said in the heat of emotion that were then released to the press by the "dissident majority," after Kaysen had made clear his intention to ignore the faculty's vote—Gödel overcame his reserve and spoke up, delivering himself in cool and rational terms. (His allies were relieved at his consistent reasonableness throughout his public remarks; in a pow-wow before the

12 Weil had come to the Institute from France and had been one of the original participants in the pseudonymous existence of "Nicolas Bourbaki." It was under this name (Bourbaki was identified as "formerly of the Royal Poldavian Academy") that a group of young mathematicians published at least two dozen mathematical treatises of the highest order, bringing the level of proof up to a new standard of rigor.

faculty meeting, he had privately speculated that perhaps Bellah, who had originally come from Canada, was so favored by the board of trustees because its director had once been the Canadian ambassador and perhaps Bellah had been a spy for Canada! This, says my source for this story, who, after all these years, still wishes to remain anonymous, was typical of Gödel's reasoning. Since Gödel had read Bellah's works and found them unimpressive, he was seeking the sufficient reason that would render the authorities' decision intelligible.)

In the public meeting, Gödel distinguished, in true Gödelian fashion, the *influence* ideas may have from their objective *truth*. (Intimations of Plato, castigating the Sophists of his day.) Bellah's defenders spoke only of the former, not of the latter. Bellah's work, Gödel was reasonably pointing out, may have influenced many in his own field, but that, in itself, is no grounds for thinking them true. Fashionable ideas aren't necessarily true ideas. The counterclaim that in such fields as sociology there is nothing *but* influence to consider, the notion of objective truth being inapplicable (Kaysen, responding to Gödel, pushed this line a bit) amounted, for Gödel, to the most severe delegitimatizing possible of a field like sociology. "He [Gödel] also pointed out that many scientists of great intelligence, originality, learning, and influence have produced completely wrong theories, for example, Stahl, the inventor of the phlogiston theory."

Yet when it came to the vote, Gödel was one of the few mathematicians who didn't vote against the nomination, once again finding it impossible to defy authority. The final vote of the faculty was 13 against Bellah, 8 for, and 3 abstentions. One of the abstainers was Kurt Gödel. This was final confirmation for Gödel's mathematical colleagues that "the logic, it was just impossible."

The whole sad business of the Bellah nomination was brought to a fittingly sad conclusion: Bellah's daughter died, and Bellah, in grief, simply withdrew from consideration. Not too long after the Bellah affair, Kaysen left the Institute, exhausted by his mathematical adversaries (as Oppenheimer and Flexner had themselves resigned, utterly wasted). The current director of the Institute is said by the mathematicians to be a reasonable man.

So much passion had been generated by the Bellah affair that in speaking with some of the participants now, more than a quarter of a century later, I could still feel blasts of heat rising up out of the past. Kochen told me that during the troubles over Bellah, Gödel would sometimes call to speak to him about it. "He was very distressed at the incivility of the atmosphere."

So despite the fact that he had briefly been the ally of his mathematical colleagues at the Institute, judging the work of the proposed new member as they judged it, his isolated exile continued. In fact, if anything it deepened.

"I Can Only Make Negative Decisions"

Karl Menger, Gödel's old acquaintance from the Vienna Circle, who was happily ensconced at Notre Dame, wrote:

> On every one of my admittedly infrequent trips to Princeton, I had long talks with Gödel. Apart for his friendship with Einstein and (especially after the latter's death)

> with Morgenstern, Gödel seemed to me rather lonely. Once he asked to my surprise, "Where is Artin now?" [This is the algebraist Emil Artin.] And when I answered, "In Princeton; I spoke to him yesterday," Gödel said, "I thought he left long ago. I haven't seen him in years."

Even more sadly, Menger theorized that the isolation that Gödel experienced at the Institute contributed to his lack of publications:

> At no time in his life did Gödel need intellectual stimulation to conceive and develop original and unexpected ideas. But he needed a congenial group suggesting that he *report* his discoveries, reminding and, if necessary, gently pressing him to write them down. All this he *had* at the beginning of his stay in Princeton with regard to the publication of his two booklets and his article on Russell. And he presumably could have found such support later. But apparently he never looked for it, and no one seemed to volunteer. The fact is that I could not observe anything of the sort in the 1950's. Rather, it soon became clear to me that he wrote up many brilliant ideas only for his desk drawer if at all. From the point of view of the outside world, his incomparable talent was lying lamentably fallow.

It was in this deep isolation that the paranoid tendencies from which Gödel had suffered even in his youth took on substance. Perhaps this darkening of his mental outlook would have been inevitable with age. Still, the imposed isolation, laced with the genuine hostility of his immediate peers, certainly couldn't have been good for him. As we used to say in the sixties: just because you're paranoid doesn't mean they're not really after you.

After Einstein's death, Gödel's deepest identification appeared to have been with Leibniz. So true was this that Gödel extended his own paranoiac fantasy of imperiled rationality so that it extended to the seventeenth-century rationalist.

In Gödel's estimation Leibniz was an even greater thinker than posterity has realized and had carried his ideas for a *characteristica universalis*—or an alphabet of thought which would be used to represent thoughts in a logical way, rendering their internal logical relations transparent—to a more advanced stage than the written testimony suggests. Gödel had confided in Karl Menger his suspicions that some of Leibniz's "important writings . . . had not only failed to be published, but [had been] destroyed in manuscript."

"Who could have an interest in destroying Leibniz's writings?" Menger had queried.

"Naturally, those people who do not want men to become more intelligent," was the logician's reply.

Menger then suggested that the iconoclastic free-thinker Voltaire would be a more likely target of censorship, but Gödel disagreed:

"Who ever became more intelligent by reading Voltaire's writings?"

Menger mentioned the interchange to Oskar Morgenstern, who had something of his own to relate on the subject of Leibniz and Gödel. He, too, had been alerted by Gödel as to the deliberate suppression of Leibniz's contributions and had tried to argue the logician out of his conviction. Finally, to convince Morgenstern, Gödel had taken the economist to the university's Firestone Library and gathered together "an abundance of really astonishing material," in Morgenstern's words. The material consisted of books and articles with exact references to published writings of Leibniz, on the one hand, and the very works cited, on the other. The primary sources were all missing the material that had been cited in the secondary sources.

"This material was really highly astonishing," a flabbergasted (if unconvinced) Morgenstern admitted.

Gödel had always worried that he wasn't living up to what the Institute had expected of him; this made him feel not only guilty but also insecure. Hard as it is to believe, the man who had been cited by Harvard as having produced the most important mathematical discovery of the century—the thinker who is generally pointed to as second only to Einstein in establishing the Institute as the haunt of intellectual divinities touching down briefly upon Earth—would sometimes call Morgenstern in a panic, saying that he expected to be thrown out. He also reported his suspicions that there were those who were trying to kill him, that his wife Adele had given away all his money, and that his doctors understood nothing of his case and were conspiring against him.

Oskar Morgenstern did remain a wonderful friend to Gödel, loyally devoted, Gödel's one abiding link, beside his wife Adele, to the old days in Vienna. Even when Morgenstern was dying of metastasized cancer, a fact tragically apparent to all his acquaintances but Gödel, his journal entries are filled with his concern for the logician.

> Today . . . Kurt Gödel called me again . . . and spoke to me for about 15 minutes. . . . After briefly asking how I was and asserting that . . . my cancer would not only be stopped, but recede . . . he went over to his own problem[s]. He asserted that the doctors are not telling him the truth, that they do

not want to deal with him, that he is in an emergency (exactly what he told me with the same words a few weeks ago, a few month ago, two years ago), and that I should help get him into the Princeton Hospital. . . . [He] also assured me that . . . perhaps two years ago, two . . . men appeared who pretended to be doctors . . . . They were swindlers [who] were trying to get him in the hospital . . . and he . . . had great difficulty unmasking them. . . . It is hard to describe what such a conversation . . . means for me: here is one of the most brilliant men of our century, greatly attached to me, . . . [who] is clearly mentally disturbed, suffering from some kind of paranoia, expecting help from me, . . . and I [am] unable to extend it to him. Even while I was mobile and tried to help him . . . I was unable to accomplish anything . . . . [Now,] by clinging to me—and he has nobody else, that is quite clear—he adds to the burden I am carrying.

This was the journal entry for 10 July 1977. Sixteen days later, Oskar Morgenstern was dead. Hours after the economist's death, Gödel called his house, expecting to speak with him, to pour out the content of his dark delusions. The news that his one trusted ally had just died so shocked him that Gödel simply hung up the phone without saying a word.

Adele, too, was experiencing health problems and had to be hospitalized during this period, and so Gödel was left to fend for himself for the autumn months, and then into the winter. With Morgenstern dead, and Adele away, the logician's decline was precipitous.

Perhaps the only one who tried to make contact with Gödel during these last few months of his life was the faithful acolyte/logician, Hao Wang. Wang was out of the country from mid-September to mid-November of 1977, but right before his departure he called Gödel, to tell him that he was coming by to see him. Wang came bearing a chicken that his wife had prepared for Gödel. Always a valetudinarian, excessively watchful of what went into his body, with recurring fears that he was being poisoned, Gödel's abstemiousness was now advancing to self-starvation. When Wang arrived at the house on Linden Lane, Gödel "eyed him suspiciously" and refused to open the door. Wang left the chicken on the doorstep and departed.

Wang did manage to gain entry into the Gödel house on 17 December and seems to have been reassured by Gödel's manner and presence (though he must have been emaciated). "His mind remained nimble and he did not appear sick. He said, 'I've lost the faculty for making positive decisions. I can only make negative decisions.'"

Adele returned home from her own hospital stay at the end of December and on 29 December (with the help of Hassler Whitney) persuaded Gödel to enter Princeton Hospital. "It was said that G[ödel]'s weight was down to sixty-five pounds before his death and that, toward the end, his paranoia conformed to a classic syndrome: fear of food poisoning leading to self-starvation."

Kurt Gödel died in the fetal position on Saturday, 14 January 1978, at one in the afternoon. According to the death certificate, on file in the Mercer County Courthouse in Trenton, he died of "malnutrition and inanition" caused by "personality disturbance."

Karl Menger contributed one last anecdote:

> In one of his last telephone calls before his own death (in July, 1977) Morgenstern described an event that evoked in

> me memories that long ago had somewhat estranged me from Gödel[13]—but it evoked them by its *contrast* to those memories, so that Morgenstern's story moved me very much. Once again it was a question of Gödel's rights, where his punctiliousness knew no bounds. What had happened was that Gödel, apparently suffering severely, sought and was granted admission to a Princeton hospital, but soon thereafter insisted that he had no right to one of the benefits proffered, since his insurance policy did not provide for it. He therefore refused to accept the benefit. The details of the case escape me now, though of course I am convinced that Gödel's logic in interpreting the insurance contract was superior to the hospital's. But be that as it may, in his juridical precision, Gödel unshakably maintained his ground.

Gödel was buried on 19 January in the Princeton Cemetery on Witherspoon Street; the funeral was small and private. But on 3 March there was a memorial service held at the Institute and presided over by André Weil. The speakers were Hao Wang, Hassler Whitney, and, as a last-minute fill-in for the prominent logician Robert Solovay, who had flown in from California but whose rented car had swerved into a snow ditch, Simon Kochen.

Kochen recalled in his tribute to Gödel that at his Ph.D. oral exam, his examiner, Stephen Kleene, had asked him to name five of Gödel's theorems. The point of the question was that "Each of the theorems . . [was] the beginning of a whole branch of modern mathematical logic." Proof theory, model

13 Menger is referring here to Gödel's outrage that his rights as a university Dozent had been tampered with by the Third Reich.

theory, recursion theory, set theory, intuitionist logic; all had been transformed by, or, in certain cases, had gotten their inception from, Gödel's work.

Kochen then compared Gödel's work to Einstein's, in terms of the way in which both grew out of their deep foundational thinking. "That was an obvious comparison to make," Kochen told me.

He also, far more surprisingly, compared Gödel's work to that of Kafka, not knowing that Gödel himself had been an admirer of the writer.[14] Both men combined their strongly "legalistic" bent, in Kochen's words, with an otherworldly, almost surreal, ability to create self-contained worlds, worlds that might seem at first blush to run counter to logic but which are compounded of the very stuff of logic. "There is an Alice-in-Wonderland quality to both men's work," Kochen said to me.

Kochen told me that had he had time to prepare his remarks, the analogy to Kafka would probably never have been made by him. It was because he had to come up with something quickly that he reached for the vague suggestion that Gödel's work had always impressed on him, but that he had never bothered to think out. But everyone I spoke with who had been present at the memorial service remembers Kochen's remark. The comparison with Kafka, another sui generis mind out of Central Europe, managed to capture something startling and true.

14 Gödel wrote his mother on 4 and 19 July 1962, of his "recent discovery" of the "modern poet," Franz Kafka. Gödel also enjoyed abstract and surrealist painting. In other respects, his cultural tastes are notable for their childlike quality. Not only did he prefer fairy tales to Goethe and Shakespeare, writing to his mother that only in such tales is the world represented as it ought to be, but he was also a great fan of Disney movies and saw *Snow White* at least three times.

Incompleteness (All Over Again)

Einstein and Gödel shared, along with so much else, a preoccupation with the nature of time. Despite the popular distortions, to a certain extent encouraged by the vague suggestions of the word "relativity," Einstein was, as we have already seen, as far from interpreting his famous theory in subjective terms as it is possible to be. On the contrary, on his interpretation,

Einstein and Gödel walking on the suppressed path that went from Fuld Hall to Olden Farm.

relativity theory offers a realist description of time that is startlingly distinct from our subjective experience of time. The great yawning chasm between the "out yonder" and the "in here" is stretched even wider, on the Einsteinian hypothesis, since objective time—the time that is described in the equations of relativity theory—is lacking the very feature that seems to provide the essential stab to our subjective experience of time: its inexorable flow, ultimately lighting all our yesterdays the way to dusty death.[15] Is there anything we know more intimately than the fleetingness of time, the transience of each and every moment?

Yet, strangely enough, it isn't so . . . not if we take Einstein's physics seriously. The nature of reality that spills forth from Einstein's physics is so much more startling than the simplistic, undergraduate-beloved shibboleth: everything is relative to subjective points of view. In Einstein's physics, there is no passage of time, no unidirectional flow away from the fixed past and toward the uncertain future. The temporal compo-

15 An Op-ed piece in *The New York Times*, which ran, appropriately enough, on New Year's Day, beautifully delineated the chasm between subjective and objective time: "A hundred years ago today, the discovery of special relativity was still 18 months away, and science still embraced the Newtonian description of time. Now, however, modern physics' notion of time is clearly at odds with the one most of us have internalized. Einstein greeted the failure of science to confirm the familiar experience of time with 'painful but inevitable resignation.' The developments since his era have only widened the disparity between common experience and scientific knowledge. Most physicists cope with this disparity by compartmentalizing: there's time as understood scientifically, and then there's time as experienced intuitively. For decades, I've struggled to bring my experience closer to my understanding."

nent of space-time is as static as the spatial components; physical time is as still as physical space. It is all laid out, the whole spread of events, in the tenseless four-dimensional space-time manifold. The distinctions we make between the past and the present and the future—distinctions that are so emotionally fraught and without which we can't even begin to describe our inner worlds—only have relevance *within* those inner worlds. Objective time, as it is characterized in relativity, can't support the distinction between the past and present and future. Or, as Einstein told Rudolf Carnap, "the experience of the now means something special for man, something essentially different from the past and the future, but this important difference does not and cannot occur within physics."

Understanding relativity theory to imply that there is no absolute *now* flowing along on a relentless tide of temporality, Einstein, living "under the sword of Damocles," seemed to take comfort in his vision of tenseless physical objectivity. In a condolence letter to the widow of Michele Besso, his longtime friend and fellow physicist, Einstein wrote: "In quitting this strange world he has once again preceded me by just a little. That doesn't mean anything. For us convinced physicists the distinction between past, present, and future is only an illusion, albeit a persistent one."

It is a vision of impersonal objectivity sufficient to extract the bitterness, at least for Einstein, from the thought of one's own personal demise, than which there are few thoughts more unpalatable. Einstein's imperturbability recalls, in its transcendence, the death of Socrates that had so inspired Plato and, through Plato, all of Western civilization. This is scientific realism carried to heroic heights. The physicist who discussed the meaning of time on his daily walks with the logician was dying, and he knew it.

Gödel, no less than Einstein, believed that time is nothing like what it *seems* to us to be. His personality may not have been of the sort to allow him to use his vision of time to transcend the fears—both real and imagined—that tormented his mortal existence; but nonetheless it did, perhaps, offer him some degree of comfort. His own work on relativity theory had provided him a model of time that seemed to have appealed to him on a deep level, to mesh with the very substance of the man, as his embrace of Platonism had done.

Gödel, of course, had a long-standing interest in physics. He had first entered the University of Vienna intending to study physics and did so for the first two or three years while a student there, before switching to math. His relationship with Einstein rekindled his earlier interest in physics, and at some point in their relationship Gödel began to ponder relativity theory for himself. He came up with an entirely unique model satisfying Einstein's field equations in general relativity, a model as Alice-in-Wonderland-like as anything else he had ever done.

In Gödel's model, time is cyclical. Not only are all events laid out in indifference to tensed distinctions between past, present, and future, but also endless repetitions of the patterns occur, and the parallelism between space and time, implicit in relativity theory, is extended further. "It turns out," wrote Gödel, "that temporal conditions in these universes show . . . surprising features, strengthening the idealistic viewpoint (according to which all change is actually an illusion, nonobjective). Namely, by making a round trip on a rocket ship in a sufficiently wide curve, it is possible in these worlds to travel into any region of the past, present, and future, and back again, exactly as it is possible in other worlds to travel to distant parts of space."

Gödel published his solution to Einstein's equations in the *Festschrift* volume in honor of Einstein's seventieth birthday.[16] Einstein's published remarks on the paper, also published in the *Festschrift*, acknowledge having been "disturbed" by the possibility of looping timelike lines, allowing one to return to the past, that Gödel gleefully expounded. Einstein's response both pays tribute to the validity of Gödel's deductions while also suggesting that Gödel's solution might "be excluded on physical grounds."

It is unclear how much of his cosmological work Gödel had shared with his daily walking partner before handing him over the results on his seventieth birthday; but Einstein's reaction to the paper suggests that Gödel had shared little. Gödel's closed loops of time, allowing one, at least theoretically, to return to the past, were accepted by Einstein as formally possible, in the sense that Gödel had shown that this model of time solves Einstein's field equations. But as a physicist and a man of common sense, Einstein would have preferred that his field equations excluded such an Alice-in-Wonderland possibility as looping time.

16 Schilpp had first approached Gödel in 1946 to contribute an essay for this *Festschrift* in honor of Einstein. Gödel immediately agreed, but there were to be many postponements of delivery of the final paper. Schilpp had hoped to have his volume ready for Einstein's seventieth birthday (14 March 1949). Gödel did not finish his article until a month before and even then held onto it. Schilpp got Gödel to agree to present Einstein with the paper at the gala birthday celebration thrown for him at the Princeton Inn on 19 March. Soon after, Schilpp received a copy of the paper.

But the model of cyclical time seemed to have appealed to Gödel very much. Did Gödel actually like the idea of being able to go back and live his life all over again? Did he, too, like his friend Einstein, draw some sort of solace from his contemplation of the real nature of time, distinct from the unidirectional finality of our experience of it?

Who knows? The opacity of the logician prevails. However, there is one interesting fact that perhaps allows us a peek behind the opacity. Gödel took his quite extraordinary solution to Einstein's equations so seriously that he descended, probably for the only time in his life, from the highest reaches of Reine Vernunft to try to acquire actual empirical (!) data to support his closed-looped model for time. John Archibald Wheeler and Kip Thorne, two of the most prominent physicists of their days, who had collaborated (with Charles Misner) on a marvelous book on gravitation, were closely questioned by Gödel in the early 1970s as to whether they had found any evidence for, or against, a preferred sense of rotation of the galaxies. Gödel was clearly disappointed in them, Wheeler reported, when they confessed that they just hadn't looked into the question:

> It turned out that he himself, as a preliminary step to get some evidence, had taken down the great Hubble atlas of the galaxies. Gödel, whom you think of as the mathematician among mathematicians, had taken a ruler and got the angle and made a statistics of these numbers and concluded that within the statistical error there was no preferred sense of rotation. . . .
>
> About a year after our visit to Gödel I was down the hall here in the office of Jim Peebles [a prominent Princeton

astrophysicist] talking to him about cosmology, and a student came in and threw down on the table a big thing. "Here it is, Professor Peebles!" So I said to him, "What is it?" He said, "It's my thesis." "What's it about?" "It's about whether there is any preferred sense of rotation in the galaxies." "How marvelous," I said, "Gödel will be *so* pleased." "Who is Gödel?" "Well," I said, "if you called him the greatest logician since Aristotle you'd be downgrading him." "Are you kidding?" "No, no." "What country does he live in?" "Right here in Princeton," I answered. So I picked up the phone and dialed Gödel, reached him at home, and told him about this. Pretty soon his questions got to the point I couldn't answer them. I turned it over to the student, and pretty soon it got to the point that the student couldn't answer them. He gave the phone to Peebles, and when Peebles finally hung up he said, "My, I wish we talked to Gödel before we did the work."

When Gödel presented his ideas on relativity at the Institute, which took place several years before the conversations with Wheeler, Thorne, and Peebles, the physicists present all expressed astonishment at how well the mathematical logician had grasped all the subtleties of the physical theory. But of course he had had the privilege of discussing the intricacies of the theory with the theoretician himself—even as the theoretician confessed that, at the end, he only went to his office to have the privilege of walking home each day with the logician, the two great minds of the twentieth century able to share, at least for a while, their intellectual exile with one another.

It is tempting to connect Gödel's attraction to these closed time loops with a passing remark that Hao Wang made, indicating how Gödel had thought of his life as incomplete:

> In philosophy Gödel has never arrived at what he looked for: to arrive at a new view of the world, its basic constituents, and the rules of their composition. Several philosophers, in particular Plato and Descartes, claim to have had at certain moments in their lives an intuitive view of this kind totally different from the everyday view of the world.

And, again, Wang made reference to a transcendental experience that Gödel had awaited all his life:

> He also looked for (but failed to obtain) an epiphany (a revelation or sudden illumination) that would enable him to see the world in a different light. (In his conversations with me, he repeatedly said that Plato, Descartes, and Husserl all had such an experience.)

Philosophy had inspired Kurt Gödel's formidable mathematical career from the beginning. It had been his focus ever since his first course at the University of Vienna in the history of philosophy, when Gödel, like so many lovers of abstraction, had found in Plato a vision of reality that answered to his intellectual love. As philosophy had been his end, so, too, it was by philosophy's light that he judged his life, finally, incomplete. No longer believing that it was possible to change other people's minds, not even by way of a priori proof, he awaited the epiphany that would change his own. With the sense of his own incompleteness—and perhaps, too, with the preserved death-terror of a child believing that his eight-year-old heart had been fatally damaged—he was drawn to his model of an eternal life of cyclical time, a model which undermines the reality of personal death.

If time loops back on itself, as Gödel in the tormented last years of his life sought empirically to corroborate, then a young Gödel will once again sit in a college classroom in Vienna, transfigured by the notion of the infinite eternal verities lying suspended beyond all the human imperfections, the confoundments and obfuscations and distortions that make him wonder how people can ever understand one another at all. And he will think about using the language of mathematics in a way that no one has thought to use it before, so that it can talk about itself—only precisely, because mathematically, so that everyone will understand. He will dream, silently and audaciously, of proving a mathematical theorem the likes of which has never before been seen, a mathematical theorem that will illuminate the nature of mathematics itself.

And then he will do it.

Notes

Introduction

p. 13 "Hitler shakes the tree": Quote unattributed, in "Bad Days on Mount Olympus: The Big Shoot-Out in Princeton," *Atlantic Monthly*, February 1974.

p. 19 "after its driver": Helen Dukas, Einstein's secretary, in a letter to C. Seelig as quoted in Abraham Pais, *"Subtle is the Lord . . .": The Science and the Life of Albert Einstein* (Oxford: Oxford University Press, 1982), p. 473.

p. 20 All of his thinking: Harry Woolf, ed., *Some Strangeness of Proportion: A Centennial Symposium to Celebrate the Achievements of Albert Einstein* (Reading, MA.: Addison-Wesley, 1980), p. 485.

p. 23 The *Encyclopedia of Philosophy*'s article "Gödel's Theorem": J. van Heijenoort, *The Encyclopedia of Philosophy*, Vol 3, ed., Paul Edwards (New York: Macmillan Publishing Co., Inc., and the Free Press, 1967), pp. 348–9.

p. 24 "He is the devil": David Foster Wallace, "Approaching Infinity," *Boston Globe*, 12 December 2003.

p. 29 "Science without epistemology": Paul A. Schilpp, *Albert Einstein, Philosopher-Scientist* (New York: Tudor, 1949), p. 684.

p. 29 "No story of Einstein": Woolf, op. cit.

p. 31 "the logic of Aristotle's successor": Armand Borel, "The School of Mathematics at the Institute for Advanced Study," in *A Century of Mathematics in America*, ed. Peter Duren (Providence, RI: American Mathematical Society, 1989), p. 130.

p. 31 "I don't believe in natural science": Quoted in Ed Regis, *Who Got Einstein's Office? Eccentricity and Genius at the Institute for Advanced Study* (Cambridge, MA: Perseus, 1987), p. 58. I also corroborated the story with John Bahcall.

p. 32 "After that . . . I just gave up": Private conversation, May 2002.

p. 32 "the laws of nature are a priori": Reported to me by Paul Benacerraf, who was told of the conversation by Chomsky.

p. 33 "*um das Privileg zu haben*": In a letter to Bruno Kreisky (Bundesmeister für Auswartige Angelegenheiten of Austria) dated 25 October 1965.

p. 34 "two-membered 'natural kind' ": Hao Wang, *Reflections on Kurt Gödel* (Cambridge, MA: MIT Press, 1987), p. 2.

p. 37 "It works, yes": Michael Frayn, *Copenhagen* (New York: Anchor Books, 1998), pp. 71–2.

p. 39 "Gödel's findings seem to have even more far-reaching consequences": William Barrett, *Irrational Man: A Study in Existentialist Philosphy* (New York: Anchor Books, 1962, 1990), p. 39.

p. 42 "the religious paradise of youth": Schilpp, op. cit., p. 5.

Chapter 1: A Platonist among the Platonists

p. 54 a brief "History of the Gödel Family": Reprinted in *Gödel Remembered: Salzburg 10–12 July 1983*, ed. P. Weingartner and L. Schmetterer (Napoli: Bibiopolis, 1987).

p. 58 John Dawson, *Logical Dilemmas: The Life and Work of Kurt Gödel* (Wellesley, MA: A K Peters LTD, 1997), p. 14.

p. 58 "an exile in Czechoslovakia": Dawson 1997, op. cit., p. 15.

p. 58 "his interest in precision": Wang 1987, op. cit., p. 41.

p. 60 until John Dawson undertook the formidable task: See also Dawson's (1997) excellent biography of Gödel that has already been cited.

p. 60 manuscripts for articles that he had promised to deliver: For example, the article for Rudolf Carnap's *Festschrift*, which he had promised to P. A. Schilpp in 1953. Gödel produced no less than six drafts of the "brief note" which grew into a lengthy manuscript. He kept revising until 1959, when he finaly wrote to Schilpp, calling it quits. Two of the six versons were posthumously published under the title "Is Mathematics the Syntax of Language?" in volume III of Gödel's *Collected Works*.

p. 60 a hysteria of discretion: Solomon Feferman expressed this double aspect of Gödel's personality this way: "What I find striking here is the contrast on the one hand between the depth of Gödel's convictions which underlay his work, combined with his sureness of insight leading him to the core of each problem, and on the other hand the tight rein he placed on the expression of his true thoughts." See Solomon Feferman, "Kurt Gödel: Conviction and Caution," in *Gödel's Theorem in Focus*, ed. S. G. Shanker (London: Croom Helm, 1988), p. 111.

p. 63 I remember my own: See Rebecca Goldstein, "Writers on Writing: On the Wings of Enchantment," *The New York Times*, 19 December 2002.

p. 63 "Here is the life, Socrates": *Symposium*, 211d-211e. Translated by William S. Cobb (State University of New York, 1993).

p. 64 Gödel had been drawn toward number theory: Wang, 1987, op. cit., p. 22.

p. 66 "It is a measure of Gödel's status": Jaakko Hintikka, *On Gödel* (Belmont, CA: Wadsworth Thomson Learning, 2000), p. 1.

p. 72 "Theory of Café Central": Quoted in Alan S. Janik and Hans Veigl, *Wittgenstein in Vienna: A Biographical Excursion Through the City and its History* (New York: Springer, 1998), p. 88.

p. 74 The philosopher Karl Popper . . . waited with impatience: See David Edmonds and John Edinow, *Wittgenstein's Poker: The Story of a Ten-Minute Argument Between Two Great Philosophers* (New York: HarperCollins, 2001), for a lively discussion of Popper's exclusion from the Vienna Circle and the antagonism to Wittgenstein that lay behind it.

p. 74 "The Circle's voice," Edmonds and Edinow, op. cit., p. 163.

p. 75 He was dismayed: He himself acknowledged in a letter to the mathematician Menger that he had contributed (presumably because of his disinclination to address it) to the misapprehension: "In conseq[uence] of frequent tiredness I hardly ever answer letters before a week's time. But in this case there was moreover a sp[ecial] reason, namely that I always have inhibitions in writing about my relationship to the Vienna Circle *because* I never was a logical pos[itivist] in which the term is commonly understood and explained in the [manifesto of] 1929. On the other hand by some publ[ications] (probably in part through my own fault) the impr[ession] is created that I was." Quoted in Wang 1987, op. cit., p. 49.

p. 80 "The pleasant atmosphere": Rudolf Carnap, "Intellectual Autobiography," in *The Philosophy of Rudolf Carnap*, ed. Paul A. Schilpp (La Salle, IL: Open Court, 1963).

p. 80 footnote 7: Boltzmann had succeeded: For a fascinating discussion of Boltzmann's life and works see David Lindley, *Boltzmann's Atom: The Great Debate That Launched A Revolution in Physics* (New York: Free Press, 2001).

p. 81 "especially interested in the formal-logical": Herbert Feigl, "The Wiener Kreis in America," in *The Intellectual Migration: Europe and America, 1930–1960*, ed. Donald Fleming and Bernard Bailyn (Cambridge, MA: Harvard Univeristy Press, 1969), p. 635.

p. 81 "that almost seemed to exude sincerity": Karl Menger, *Reminiscences of the Vienna Circle and the Mathematical Colloquia*, ed. Louise Golland, Brian McGuinness, and Abe Sklar (Dordrecht: Kluwer, 1994), p. 63.

p. 81 When anything came up in conversation: Menger, op. cit., p. 64.

p. 82 "Schlick especially seemed to resent this": Menger, op. cit., p. 65.

p. 84 "Uncritical, run-down aristocrats": Menger, op. cit., pp. 61–2.

p. 85 "our philosophical movement with its international trade name": Feigl, op. cit., p. 630.

p. 85 "The truths of pure mathematics": Feigl, op. cit., p. 652.

p. 88 "a rather dingy room," Menger, op. cit., p. 55.

p. 90 "the Austrian equivalent": Ray Monk, *Ludwig Wittgenstein: The Duty of Genius* (New York: Penguin, 1990), p. 72.

p. 94 "We were both cross from the heat": Bertrand Russell to Ottoline Morrel, 27 May 1913.

p. 95 footnote 13: "were taken from authors": Allan Janik and Stephan Toulmin, *Wittgenstein's Vienna* (New York: Simon and Schuster, 1973), p. 27.

p. 95 Partly it was the Viennese aspect in his thinking: This is a dominant theme in Janik and Toulmin, op. cit. "Those of us who attended his lectures [at Cambridge University] . . . still found ourselves looking upon his ideas, his methods of argument and his very topics of discussion as something totally original and his own. . . . If there was an intellectual gulf between him and us, it was not because his philosophical methods, style of exposition and subject matter were (as we supposed) unique and unparalleled. It was a sign, rather, of a culture clash: the clash between a Viennese thinker whose intellectual problems and personal attitudes alike had been formed in the neo-Kantian environment of pre-1914, in which logic and ethics were essentially bound up with each other and with the critique of language (*Sprachkritik*), and an audience of students whose philosophical questions had been shaped by the neo-Humean . . . empiricism of Moore, Russell and their colleagues" (pp. 21–2).

p. 95 "a quintessentially Viennese figure": Monk, op. cit., p. 20.

p. 96 pronounced a genius by Russell: "Wittgenstein's recurring thoughts of suicide between 1903 and 1912, and the fact that these thoughts abated only after Russell's recognition of his genius, suggest that he accepted this imperative (Weininger's) in all its terrifying severity." Monk, op. cit., p. 25.

p. 99 "stranger than others by orders of magnitude": Jaakko Hintikka, op. cit, p. 3. Hintikka is here speaking even more widely than of logic; he is speaking of all of mathematics.

p. 101 "We shall see contradiction": *Remarks on the Foundations of Mathematics* (Cambridge, MA: MIT Press, 1967), p. 110.

p. 101 the later Wittgenstein came to regard the entire field as a "curse": For example, he writes: "The curse of the invasion of mathematics by mathematical logic is that now any proposition can be represented in a mathematical symbolism, and this makes us feel obliged to understand it. Although of course this method of writing is nothing but the translation of vague ordinary prose." Wittgenstein, op. cit., 155.

p. 103 footnote 16: "an Indian poet much in vogue": Monk, op. cit., p. 243.

p. 104 footnote 17: "Although Gödel had not persuaded Carnap on this fundamental issue": Eckehart Köhler, "Gödel and the Vienna Circle: Platonism versus Formalism," in *History of Logic, Methodology, and Philosophy of Science,* section 13 (Vienna Institute for Advanced Studies). Later cited in S. G. Shanker, ed., *Gödel's Theorem in Focus* (London: Croom Helm, 1988).

p. 105 "I shall relate to you": Quoted in Janik and Veigl, op. cit. p. 63.

p. 105 "a mythological character": Monk, op. cit., p. 284.

p. 105 "Schlick adored him": Feigl, op. cit., p. 638.

p. 105 "Feigl had always had an unusual ability to get along with everyone": Menger, op. cit., p. 66.

p. 105 "limitless admiration for Carnap": Ibid.

p. 106 "I once wanted to give a few words in the foreword": Letter to Ludwig von Ficker, quoted in Monk, op. cit., p. 178. The letter is undated, but Monk says it was almost certainly written 19 November 1919.

p. 110 "a slim, unusually quiet young man": Menger, op. cit., p. 201.

p. 110 "a very unassuming, diligent worker": Feigl, op. cit., p. 640.

p. 111 "Some reductionism is correct": Hao Wang, *A Logical Journey: From Gödel to Philosophy* (Cambridge, MA: MIT Press, 1996).

p. 113 "[A] friend described going to Beethoven's door": Bertrand Russell to Ottoline Morrell, 23 April 1912.

p. 114 "I had my tonsils out": Fania Pascal, "Wittgenstein: A Personal Memoir," in *Recollections of Wittgenstein*, ed. R. Rhees (Oxford: Oxford Press, 1984), pp. 28–9.

p. 117 "In the early 1970's": Menger, op. cit., p. 230.

p. 117 " '*logische Kunststücke*' ": The relevant passage is in *Remarks on the Foundations of Mathematics,* Appendix I, p. 19. "You say: . . . , so *P* is true and unprovable. That presumably means: Therefore *P*. That is all right with me—but for what purpose do you write down this 'assertion'? (It is as if someone had extracted from certain principles about natural forms and architectural style the idea that on

Mount Everest, where no one can live, there belonged a chalet in the Baroque style. And how could you make the truth of the assertion plausible to me, since you can make no use of it except to do these little conjuring tricks?"

p. 118 "As far as my theorems about undecidable propositions are concerned": Menger, op. cit., p. 231.

p. 118 "Wittgenstein's views on mathematical logic": Georg Kreisel, "Wittgenstein's 'Remarks on the Foundations of Mathematics,'" *British Journal for the Philosophy of Science*, IV (1958), pp. 143–4.

Chapter 2: Hilbert and the Formalists

p. 124 "So the geometrical figures are signs or mnemonic symbols": David Hilbert, "Mathematische Probleme. Vortrag, gehalten auf dem internationalen Mathematischer-Kongress zu Paris 1900." *Nachrichten von der Könglichen Gesellschaft der Wissenschaften zu Göttingen*, 253–97. English translation in Felix Browder, ed., "Mathematical Developments Arising from the Hilbert Problems," *Proceedings of Symposia in Pure Mathematics* XXVIII, parts 1 and 2. (Providence, RI: American Mathematical Society, 1976).

p. 128 "In mathematics we must always strive after a system": Gottlob Frege, *Begriffsschrift, eine der arithmetischen nachgebildete Normalsprache des reinen Denkens* (Halle: Nebert, 1879). English translation in Jean van Heijenoort, ed., *From Frege to Gödel: A Source Book in Mathematical Logic, 1879–1931* (Cambridge, MA: Harvard Univeristy Press, 1967), p. 279.

p. 128 "no science can be so enveloped in obscurity as mathematics": Jean Heijenoort, ed., op. cit., p. 242.

p. 136 "meaningless marks on paper": Quoted in John de Pillis and Nick Rose, *Mathematical Maxims and Minims* (Raleigh, NC, 1988).

p. 142 "Admittedly, the present state of affairs": David Hilbert, "On the Infinite," in *Philosophy of Mathematics*, ed, Paul Benacerraf and Hilary Putnam (Englewood Cliffs, NJ: Prentice-Hall, 1964), p. 141. A translation of a talk delivered 4 June 1925 before a congress of the Westphalian Mathematical Society in Münster, in honor of

Karl Weierstrass. Translated by Erna Putnam and Gerald J. Massey from *Mathematische Annalen* (Berlin) no. 95 (1925), pp. 161–90.

Chapter 3: The Proof of Incompleteness

p. 149 "the application of the Verification Principle to mathematics": Monk, op. cit., p. 295.

p. 154 footnote 2: "The completeness theorem, mathematically, is indeed an almost trivial consequence": Hao Wang, *From Mathematics to Philosophy* (New York: Humanities Press, 1974), pp. 8–9.

p. 158 No mention of Gödel . . . by Hans Reichenbach: Reichenbach's account was published in *Die Naturwissenschaften,* Vol. 18 (1930), 1093–4.

p. 160 Gödel didn't fully prove his second incompleteness theorem until after the conference: John Dawson, "The Reception of Gödel's Incompleteness Theorems," in *Gödel's Theorem in Focus,* ed. S. G. Shanker (London: Croom Helm, 1988), p. 91, footnote 2.

p. 160 How, given Gödel's *Entdeckungen,* could he not have questioned his former thinking?: See Dawson (1988) for a thorough discussion of the reaction, or initial lack thereof, to Gödel's first incompleteness theorem.

p. 161 "In science . . . novelty emerges": Kuhn, op cit., p. 64.

p. 179 What we use next is something called the diagonal lemma: See Hintikka, op. cit., p. 33.

p. 185 "Operating with the infinite can be made certain only by the finite": Hilbert 1964, op. cit., p. 151.

p. 188 The last article that Gödel was to publish in his life: "Über eine bisher noch nicht benutzte Erweiterung des finiten Standpunktes," *Dialectica* 12 (1958), pp. 280–7.

p. 188 "doubtful . . . about the completeness of the formal systems": In a letter to Constance Reid Bernays, 3 August 1966; quoted in Dawson 1997, p. 72.

p. 189 "Mathematics cannot be incomplete": Wittgenstein 1967, op. cit., p. 158.

p. 190 "No calculus can decide a philosophical problem": *Philosophical Remarks,* p. 296.

p. 190 "My task is not to talk about Gödel's proof": *Remarks on the Foundations of Mathematics* V, p. 16.

p. 191 "They are what is mystical": *Tractatus*, 6.522.

p. 195 footnote 9: how Gödel "entered my intellectual life": Stephen C. Kleene, "Gödel's Impression on Students of Logic in the 1930's," in *Gödel Remembered*, ed. Paul Weingartner and Leopold Schmetterer (Naples: Bibliopolis, 1987) p. 52.

p. 196 footnote 10: "It is very queer": "Wittgenstein's Lectures on the Foundations of Mathematics: Cambridge 1939," from the *Notes of R. G. Bosanquet, Norma Malcolm, Rush Rhees, Yorick Smythies*, ed. Cora Diamond (Ithaca, NY: Cornell University Press, 1976), Lecture XXI, pp. 206–7.

p. 200 "Gödel's theorem seems to me to prove that Mechanism is false": J. R. Lucas, "Minds, Machines, and Gödel," *Philosophy*, XXXVI (1961), p. 112.

p. 201 "What did Gödel's theorem achieve?": Roger Penrose, *Shadows of the Mind: A Search for the Missing Science of Consciousness* (Oxford: Oxford University Press, 1994), pp. 64–5.

p. 203 "Either the human mind surpasses all machines": Wang 1974, op. cit., p. 324.

p. 205 "Delusions may be *systematized*": Shervert H. Frazier and Arthur C. Carr, *Introduction to Psychopathology* (Jason Aronson, 1983), p. 106.

p. 205 "A paranoid person is irrationally rational": James W. Anderson, Associate Professor of Clinical Psychology, Northwestern University, personal communication, 7 October 2003.

Chapter 4: Gödel's Incompleteness

p. 210 He mentioned to . . . Morton White: Private conversation with Morton White, May 2002.

p. 213 footnote 3: according to Hao Wang . . . in 1970: Wang 1987, op. cit., p. 9.

p. 213 "Gödel would probably have published more": Wang 1987, op. cit., p. 29.

p. 216 revised and expanded for this volume: The original appeared in the *American Mathematical Monthly* 54 (1947), pp. 515–25. The original had been published before Paul Cohen had proved that the continuum hypothesis could not be deduced from the axioms of set theory.

p. 220 "Gödel was more withdrawn after his return from America": Menger, op. cit., p. 205.

p. 223 "How can any of us be called professor when Gödel is not?": Stanislaw M. Ulam, *Adventures of a Mathematician* (New York: Charles Scribner's Sons, 1976), p. 80. As Dawson remarks (Dawson 1997, p. 302, note 462), it "is worth noting that Gödel himself seems never to have complained about his status, either publicly or in private remarks or correspondence."

p. 223 "G married Adele Porkert on 20 September 1938": Wang 1988, op. cit., p. 47, footnote 7.

p. 226 "He had complained about the revocation of his dozentship": Menger, op.cit., p. 123.

p. 227 "And what brings you to America, Herr Bergmann?": Dawson 1997, op. cit., p. 90.

p. 227 "During the summer I heard nothing from Gödel": Menger, op. cit., p. 124.

p. 230 "His case could hardly create a precedent": Dawson 1997, op. cit., p. 148.

p. 241 Nonstandard analysis, he said, was not "a fad": Dawson 1997, op. cit., p. 244.

p. 244 "He [Gödel] also pointed out that many scientists of great intelligence": Morton Gabriel White, *A Philosopher's Story* (University Park, PA: Pennsylvania State University Press, 1999), p. 303.

p. 245 "On every one of my admittedly infrequent trips to Princeton": Menger, op. cit., p. 226.

p. 247 "Who could have an interest in destroying Leibniz's writings?": Menger, op. cit., p. 19.

p. 248 He also reported his suspicions that there were those who were trying to kill him: Dawson 1997, op. cit., pp. 249–50.

p. 248 "Today . . . Kurt Gödel called me": Morgenstern papers, Perkins Memorial Library, Duke University, folder "Gödel, Kurt, 1974–1977."

p. 250 "eyed him suspiciously": Wang 1987, op. cit., p. 133.

p. 250 "I've lost the faculty for making positive decisions": ibid.

p. 250 "It was said that G[ödel]'s weight was down to sixty-five pounds": ibid.

p. 254 footnote 15: "The Time We Thought We Knew," Brian Greene, Op/Ed, *The New York Times,* 1 January 2004.

p. 256 "temporal conditions in these universes show . . . surprising features": Kurt Gödel, "A Remark About the Relationship Between Relativity Theory and Idealistic Philosophy," in *Albert Einstein: Philosopher Scientist*, ed. Paul Arthur Schilpp (New York: MJF Books, 1949), p. 560.

p. 257 "be excluded on physical grounds": Einstein, op. cit., p. 687–8.

p. 258 "It turned out that he himself, as a preliminary step to get some evidence, had taken down the great Hubble atlas of the galaxies": Jeremy Bernstein, *Quantum Profiles* (Princeton, NJ: Princeton University Press, 1991), pp. 140–1.

p. 260 "In philosophy Gödel has never arrived at what he looked for": Wang 1987, op. cit., p. 46.

p. 260 "He also looked for (but failed to obtain) an epiphany": Wang 1987, op. cit., p. 196.

Printed in the USA
CPSIA information can be obtained
at www.ICGtesting.com
JSHW031707191023
50455JS00025B/405